Plant Health Management for Food Security
Issues and Approaches

— Editors —

Dr Gururaj Katti

Dr Anitha Kodaru

Dr Nethi Somasekhar

Dr G S Laha

Dr B Sarath Babu

Dr K S Varaprasad

2016

Daya Publishing House®

A Division of

Astral International Pvt. Ltd.

New Delhi – 110 002

ISBN: 9789351309734 (International Edition)

Published by	: **Daya Publishing House®** *A Division of* **Astral International Pvt. Ltd.** – ISO 9001:2008 Certified Company – 4760-61/23, Ansari Road, Darya Ganj New Delhi-110 002 Ph. 011-43549197, 23278134 E-mail: info@astralint.com Website: www.astralint.com
Laser Typesetting	: **Classic Computer Services**, Delhi - 110 035
Printed at	: **Replika Press Pvt. Ltd.**

डॉ एन. के. कृष्ण कुमार
उप महानिदेशक (बाग. वि.)
Dr N K Krishna Kumar
DEPUTY DIRECTOR GENERAL (Hort. Sci.)

भारतीय कृषि अनुसंधान परिषद
कृषि अनुसंधान भवन-II
पूसा, नई दिल्ली -110 012

INDIAN COUNCIL OF AGRICULTURAL RESEARCH
KRISHI ANUSANDHAN BHAWAN-II
PUSA, NEW DELHI-110 012

Foreword

The International Conference on Plant Health Management for Food security (ICPHM) held at Hyderabad from November 28-30, 2012 was a unique global event organized by the Plant Protection Association of India (PPAI), Hyderabad, India. The event brought together a large number of scientists and plant protection specialists representing different national and international as well as private and public institutions related to plant health management. Kudos to the efforts of PPAI for undertaking such a monumental task, ably supported by the Indian Council of Agricultural Research, The Centre for Agriculture and Bioscience International, International Crops Research Institute for the Semi-Arid Tropics, National Institute of Plant Health Management, Acharya NG Ranga Agricultural University and Crop Protection Federation. The main objective of the conference was to seek new global trends and directions in plant health management for food security.

Agriculture is a key sector for the economy of many countries and it can be potentially exposed to dangers which can have significant economic consequences for food, feed, and fiber sectors. Intentional introduction of certain microbes may have serious consequences on human health when food reserves get contaminated after harvesting and/or processing. Prevention and preparedness are the two basic approaches to maximize food security against any sort of calamity whether natural or manmade. Concerns of global climate change, loss of crop biodiversity and broad international consensus on policy issues of plant health management are the constant challenges to be tackled in future.

Hence, the organization of an International gathering involving key stakeholders was very timely and meticulous as it covered a wide range of topics under four main themes viz; Frontier areas of research, Pest management issues, Climate change and global warming and Stakeholder and policy issues, related to plant health management for food security. In this context, this publication of compilation on lead talks presented in the conference is a welcome step and deserves warm appreciation. This edited volume will serve the interest of various stakeholders particularly the young researchers taking forward the message from the International conference that emerging science and technology accomplishments are the key drivers for finding solutions through development of better plant health management strategies for food security.

The editorial team of this publication and the stakeholders involved in successful organization of the International conference deserve full compliments.

N K Krishna Kumar
Deputy Director General (Horticulture)
ICAR

Preface

This publication is our sincere and humble effort to compile and place before the stakeholders, the lead presentations delivered at the International Conference on Plant Health Management for Food Security organized by the Plant Protection Association of India (PPAI) at Hyderabad from November 28-30, 2012.

The publication consists of fifteen topics covering wide range of subject areas under the four main themes of the conference *viz.*, Frontier areas of research, Pest management issues, Climate change and global warming and Stakeholder and policy issues, related to plant health management for food security.

The opening chapter deals with Plant-parasitic nematodes which are one of the major emerging biotic stresses limiting crop production and causing considerable damage to global agriculture. The focus is on integration of novel research strategies with conventional ones for effective nematode management.

The second chapter broadly reviews the utility of behaviour modifying chemicals with particular reference to the role of sex pheromones in insect pest management. It also highlights successful efforts made in the development and implementation of pheromone application technology for the management of groundnut leaf miner, an important oilseed pest.

In the third chapter, the role of arbuscular mycorrhizal fungi (AMF) in improving plant growth is presented. The potential role of AMF is explored as an eco-friendly alternative approach in the biological suppression of soil borne pathogens.

The fourth chapter addresses the plant health management through using diagnostics and certification of seed material to minimize risks from pests within the country as well as from exotic and invasive pests. Emphasis is on efforts towards development of a National Plant Pests Diagnostic and certification Network.

The fifth topic provides status of the pest situation in key crops and their shifts with reference to Asia and Africa with potential action plan to update the plant protection needs. This will help in devising an action plan for the management of key pests for sustainable food security in these regions.

In the sixth chapter, application of modern tools of biotechnology for pest management is explained highlighting the need to develop scientifically sound strategies for deploying genetically engineered insect-resistant crops for sustainable crop production. Bio-safety assessment of these strategies along with conventional methods is also emphasized.

The next topic is about the abundant scope to exploit nanotechnology in plant health management. Development of smart delivery systems which facilitates enhanced use efficiency of inputs with environmental protection is shown as a distinct possibility.

The eight chapter deals with policies related to biosecurity in trade and exchange of germplasm in the context of plant health management. The need to support research, training, capacity-building, networking and information sharing activities related to biosecurity issues is amply justified for food security.

In the context of safety issues related to the role of pesticides in plant health management, pesticide residues in food have always merited serious attention. In the ninth chapter, the topic has been dealt in Indian perspective and relevant information needed for suitable interventions in pest management have been documented.

The next chapter deals with the important topics of pest forecasting and modeling in plant health management. The empirical and process-based modeling approaches adopted by researchers to provide pest forecasts are discussed.

The eleventh chapter reviews the recent trends in pesticide residue analysis through mass spectrometric methods.

In view of the environmental and health hazards posed by the toxic chemicals used for crop protection, the need for alternative ecofriendly strategies for controlling the pests is obvious. The twelfth chapter explores the potential of the novel technique of electronic beam radiation for grain disinfestation for phytosanitary purposes.

Similarly, the importance of Geographical information system (GIS) for managing plant genetic resources (PGR) with special reference to crop health management is elaborated in the next chapter. Potential uses of GIS in PGR management and Crop health management are discussed.

Entomopathogenic nematodes (EPN) have emerged as potential eco-friendly alternative agents for biological insect suppression. The fourteenth chapter examines in detail the journey of EPNs from laboratory to land in rice pest management.

In the last chapter, focus is on the Indian scenario of the invasive pest, papaya mealy bug. It details the state-wise occurrence and spread of the pest in the country and outlines the management of this pest using parasitoids.

The entire compilation has been made possible mainly through the committed and combined efforts of the lead speakers and their team after their excellent presentations during the conference. They also made our job easy by their timely submission of already well drafted manuscripts.

Last but not the least, the editing authors are highly grateful to the PPAI and all its associated members for their active guidance, cooperation and enthusiastic driving force which has enabled us to contribute towards bringing out this important publication. We sincerely hope that it will serve the purpose of providing research roadmap for young researchers in the country in the field of plant health management.

Editors

Contents

Chapter 1

Approaches in Nematode Management: New Perspectives

K S Varaprasad[1], M Nagesh[2] and B Gayatri[1]

[1]ICAR-Indian Institute of Oilseeds Research,
Hyderabad – 500 030, Telangana State
[2]ICAR-National Bureau of Agricultural Insects Resources,
Bengaluru – 560 024, Karnataka

ABSTRACT

Plant-parasitic nematodes are one of the major limiting factors in crop production causing considerable damage to global agriculture. Their ability to cause severe yield losses to several crops of economic importance necessitates the development of efficient and novel approaches to overcome the losses due to nematodes. Although chemical nematicides have a great promise for the control of these parasites, due to their hazards effects on environment, resulted in their withdrawal or restricted use. Using natural host plant resistance in management is a better option but timing and localization of the resistance response varies with the particular resistance gene and nematode interaction with host and environment. Therefore, alternative control measures are needed for overcoming stress induced by nematodes. With the advent of new era of genetics and genomics, new avenues were opened up for developing target specific strategies to combat plant parasitic nematodes. In this article, the novel management strategies against plant parasitic nematodes viz., exploitation of inherent soil antagonistic potential and soil health restoration, inducing resistance by activating plant defense systems, utilization of genomics and bioinformatics for identifying novel gene targets, RNAi silencing of specific gene products, chemical based strategies using novel active principles of plant origin and Good Agricultural Practices have been reviewed. Research focus on the above novel approaches and integration is a key for successful nematode management.

Keywords: *Plant parasitic nematodes, Genomics, Resistance, Novel approaches, Nematode management.*

Biotic stress due to insect pests, nematodes, mites and diseases is a major constraint in realizing the production and productivity in entirety. With the cultivable area shrinking coupled with population explosion, there is a need to produce potential crop yields with minimized losses due to biotic stress. Earlier Sasser and Freckman (1987) have reported 12.3 per cent average loss in crop yields on global basis in a few major crops due to plant parasitic nematodes. Global estimate of annual yield loss caused by plant parasitic nematodes is US $ 173 billions in agriculture (Elling, 2013). In India, annual crop loss due to major nematode pests is about Rs 21,068.73 million (Khan *et al.*, 2010). Development of hybrids/ improved varieties, fertigation, drip irrigation has contributed not only to increased productivity but also the biotic stresses, especially the subterranean flora and fauna. Although the modern agriculture learnt to manage the nematodes, there has been emergence of nematode menace as a serious constraint.

In recent times, emerging nematode problems that are causing serious concern include *Meloidogyne graminicola* in rice in Karnataka, Andhra Pradesh, Gujarat, Kerala, Odisha, Tamil Nadu (TN), West Bengal; *Meloidogyne triticoryzae* in rice and wheat cropping systems of north western plain; *Heterodera avenae* in wheat of Haryana and Bihar; *Meloidogyne incognita* in banana in Karnataka, Kerala, Tamil Nadu, Maharashtra; root-knot nematodes in citrus, grapevine and pomegranate in Karnataka, Maharashtra, Gujarat, Madhya Pradesh (MP); *Pratylenchus* spp. in coffee in Karnataka, TN; cyst nematode in pulses in Gujarat, MP, Maharashtra; root-knot nematode in castor, root-knot nematodes in carnation, gerbera, cherry tomato, bell-pepper, English cucumber, roses in protected polyhouses, besides the existing nematode constraints in carrot, beetsugar, ginger, potato, gherkins and vegetables. Their association with soil-borne wilt pathogens, *viz.*, *Fusarium, Verticillium, Phytophthora, Ralstonia etc.* compounds the plant damage leading to crop failures of higher magnitude.

The recent de-registration of several chemical nematicides due to their hazardous effect on environment and the impending withdrawal of methyl bromide from the pest control market necessitate the development of new methods for controlling nematode-induced crop damage. New approaches can be categorized in to soil health restoration and management; identification and utilization of inherent host plant resistance mechanisms; molecular approaches through transgenic systems, genomics, bioinformatics and novel functional pathways, RNAi technology, developing novel target-specific controls by exploiting fundamental differences between the biological processes of nematodes and their host plants; phytochemical based strategies; search for newer and novel active principles of synthetic chemicals; improvised crop and cropping related practices. The approaches can be broadly addressed under utilization of inherent soil antagonistic potential and soil health; inherent plant defense and health; genomics and molecular means; horticultural practices and GAP.

Inherent Soil Antagonistic Potential and Soil Health Management Approaches

Soils are biologically active with inherent biodiversity and intensity of biological processes and metabolisms. In simple terms, natural soils have the capability to maintain homeostasis of total biomass and biological processes that quantify their antagonistic potential and health. Due to continuous human manipulation for higher crop productivity, homeas has been destabilized in cultivated soils leading to disastrous situations. Soil health typically depends on microbiological processes such as C, N, elemental cycles, BNF in association with plant hosts, biological control processes (parasitism, pathogenesis, predation, competition, synergism, commensalism etc.), in combination with carbon recycling and organic carbon status. Soil organic carbon status and biological processes have a direct mutual relationship and role in soil health. Managing organic carbon is central to integrated soil biology management because the quantity and quality of soil organic inputs affect the activity and diversity of organisms within soil food web and they inturn, influence numerous soil properties relevant to eco system function and crop growth (Stirling, 2014). The concept of biological control in nematode management is now envisaged in a broader perspective, which covers the entire gamut of identification of biologically active soils (suppressive soils), isolation of suitable predominant antagonists, identification of their performance under different cropping conditions, development of passport data, monitoring the *in situ* behavior of the antagonists in conjunction with the behavior of nematode populations and host plant, to final availability of these antagonists in the form of suitable product, which is environmentally and economically acceptable, potentially viable and effective under different agro-ecological and storage conditions, and legally approved as a registered product.

A considerable range of fungi has been reported to colonize the eggs of cyst and root-knot nematode. *Rhopalomyces elegans* is considered to be the first discovery of a confirmed egg parasite. Among egg parasites, *Pochonia chlamydosporia* (*Verticillium chlamydosporium*), *Nematophthora gynophila* and *Paecilomyces lilacinus* have been considered important. *P. chlamydosporia* is one of the major pathogens of eggs within the cyst of *Heterodera schachtii*, *H. glycines* and *H. avenae* (Tribe, 1977). The same fungus can also parasitize the eggs of *Meloidogyne arenaria* and cause reduction in nematode population (Kerry, 1984). According to Al-Raddad (1995), When *Glomus mosseae* and *P. lilacinus* were applied separately or in combination against *M. javanica* in tomato, significant reduction in galling was observed. Similar effect was observed by Goswami *et al.* (2006) when *P. lilacinus* and *Trichoderma viride* were applied alone or in combination with mustard cake and carbofuran to manage *M. incognita* in tomato. In European countries, pre-application of *P. chlamydosporium* was found successful. *In vitro* studies were carried out by Satyandra and Nita (2010) to see the effect of various antagonistic fungi against the *M. incognita*. Among all the tested fungi, culture filtrate of *Acremonium strictum* was very effective against the nematode with regard to egg parasitism (53 per cent), egg hatching inhibition (86 per cent) and mortality (68 per cent) compared to controls. Fungi *viz.*, *Exophirala, Gliocladium, Phoma, Cephalosporium, Fusarium, Torula, Acremonium bacilosporum, Helicoon farinosum, Mortierella nana, Verticillium bulbilosum etc.*, had been reported to destroy egg shell of heteroderid nematodes. Among the nematode trapping fungi, *Arthrobotrys poligosporia, A. conoides, Dactylella oviparasitica* etc., have reached a status of commercial use on regular basis. Many bacterial antagonists parasitize or suppress plant parasitic nematodes. Some are obligate parasites like *Pasteuria* spp., while others PGPR like *Pseudomonas* spp, *Bacillus* spp., indirectly suppress nematode infection by improving plant defense systems. Ashoub *et al.* (2010) conducted *in vitro* and *in vivo* studies with different bacterial bioagents *viz., Bacillus thuringiensis, Pseudomonas fluorescens* and *Rhizobium leguminosarum* to observe their effectiveness for suppressing *M. incognita. In vitro* results showed that all biovars of *B. thuringiensis* and *P. fluorescens* besides *R. leguminosarum* caused juvenile mortality upto 100 per cent after 72 h of treating with culture filtrates. *In vivo* study exhibited that *P. fluorescens* is the most effective bacterium suppressing *M. incognita* reproduction followed by *R. leguminosarum*. In India, ICAR-Indian Institute of Horticultural Research (IIHR), Bengaluru standardized eco-friendly methods of nematode management underprotected cultivation. Nematologists at IIHR have developed successful management strategies of nematodes and other disease complexes under protected conditions using biopesticides like *P. lilacinus, Pochonia chlamydosporia, Trichoderma harzianum, T. viride* and *P. fluorescens* (Rao *et al.,* 2015).

Biopriming is a new technique of seed treatment where the seed is treated with bioagent before sowing to protect the seedling from soil borne pathogens. It may provide a better alternative for chemical seed treatment (Parvatha Reddy, 2013). More recently, two RNA viruses were found infecting *Caenorhabditis* spp where the viruses completely damaged intestinal cells of nematodes and impaired reproduction (Felix *et al.,* 2011). Since viruses have been successfully deployed as a safe and environmentally friendly means of managing insects (Hunter- Fujita *et al.,* 1998), there is possibility of utilising these viruses in nematode management. Therefore, there is an urgent necessity to catalogue, quantify, isolate, conserve and enhance the biological diversity of beneficial microbes in natural and suppressive soils establish and restore the natural biological processes and organic carbon status in cultivated soils for them to naturally control the nematodes and other soil-borne pathogens.

Inherent Host Plant Defense and Approaches to Utilize the Host Resistance

Use of nematode-resistant crop varieties have not been extensively evaluated in India or abroad, but is often viewed as the foundation of a successful integrated nematode management program. Commercially available nematode-resistant varieties are currently available only for few crops. According to All India coordinated project on nematodes, the following varieties are resistant/tolerant to nematodes in various crops (Table 1.1).

A vast plant germplasm in agri-horticultural crops still remains to be screened against nematodes for tolerance. In a resistant variety, nematodes fail to develop and reproduce normally within root tissues, allowing plants to grow and produce fruit even though nematode infection of roots occurs. Some crop yield loss can still occur however, even though the plants are damaged less and are significantly more tolerant of root-knot infection than that of a susceptible variety. A single dominant gene (subsequently referred to as the Mi or Mj gene) has been widely used in plant breeding efforts and varietal development, which confers resistance

Table 1.1: Cultivars Resistant to different Nematodes in Various Crops

Crop	Nematode	Resistant Cultivar
Tomato	Root-knot nematodes (*Meloidogyne javanica/ Meloidogyne incognita*)	PNR-7, NT-3, NT-12, Hisar, Lalit
Chilli	Root-knot nematodes (*M. javanica/M. incognita*)	NP-46A, PusaJwala, Mohini
Cowpea	Root-knot nematodes (*M. javanica/M. incognita*)	GAU-1
Mungbean	Root-knot nematodes (*M. javanica/M. incognita*)	ML-30 and ML-62
Cotton	*M. incognita*	Bikanerinerma, Sharda, Paymaster
Grapevine	Root-knot nematodes (*M. javanica/M. incognita*)	Khalili, Kishmish Beli, Banquabad, Cardinal, Early Muscat, Loose Perlett
Potato	Potato cyst nematode (*Globodera rostochiensis*)	Kufri Swarna, Kufri Giriraj

to all of the economically important species of root-knot nematode including *M. incognita, M. arenaria,* and *M. javanica.* Unfortunately, in previous research with resistant tomato varieties, the resistance has often failed as a result of the heat instability or apparent temperature sensitivity of the resistant Mi gene. In pepper, two root-knot nematode resistant varieties (Carolina Belle and Carolina Wonder) were released from the USDA Vegetable Research Laboratory for commercial seed increase in April 1997. Both varieties are open pollinated, and homozygous for the N root-knot nematode resistant gene. Preliminary research has demonstrated that these varieties confer a high degree of resistance to the root-knot nematode, however, expression of resistance is heat sensitive. Natural plant resistance is the most important attribute that is able to suppress invasion by the plant parasitic nematodes (Hussey and Janssen, 2002) but not all plant species carry resistance against nematodes or there are many crops for which appropriate resistance loci have not been identified (Roberts, 2002; Williamson and Kumar, 2006). Recombinant DNA technology allows us to transfer resistance genes (R-genes) into genome of plant to provide robust resistance against these plant parasitic nematodes. Using molecular cloning techniques several natural resistance genes against root-knot nematodes, *Meloidogyne* spp. have been identified and cloned. The first nematode resistance gene Hs1pro-1 was cloned from sugar beet (*Beta procumbens*), which showed resistance against the sugar beet cyst nematode *Heterodera schachtii. Mi* gene from tomato, confers resistance to major species of *Meloidogyne viz., M. incognita, M. javanica, M. arenaria* (Williamson, 1998; Milligan *et al.,* 1998). CaMi gene from pepper, exhibits resistance against *M. incognita* (Chen *et al.,* 2007) and *Ma* gene from the Myrobalan plum (*Prunus cerasifera*), which confers complete-spectrum, heat-stable and high-level resistance to the root-knot nematode, is remarkable in comparison with the *Mi-1* gene (Williamson and Kumar, 2006; Claverie *et al.,* 2011). Further research is necessary to characterize the usefulness of these resistant genes under different climatic conditions and utility in breeding programs.

Genetic Engineering and Genomics based Approaches

The nematode resistance in crops can be achieved in a more selective and robust manner by using genetic engineering strategies. These may include transforming crop plants with (i) Resistance genes against specific nematode pests (ii) Proteinase inhibitor genes, which affect the life processes of nematodes (iii) Lectin genes as anti-invasion and anti-migration strategies (iv) Genes for apoptosis of feeding cell and antinematode effectors (v) Genes expressing toxins against nematodes. Nematode-induced promoters are very useful for the production of sufficiently high levels of anti-nematode proteins at feeding sites by plants. Alternatively, interfering with feeding-cell development is somewhat similar to the hypersensitive response evoked by nematodes in a naturally resistant plant. Here, destruction of specific plant cells can be achieved by the localized expression of a cytotoxin such as barnase, a potent ribonuclease and plantibodies against plant parasitic nematodes. This approach, however, calls for a highly specific 'non-leaky' promoter, which is active only in the feeding cells. Another possibility is to use a two-component system, where the leakiness of the promoter in other tissues is counter balanced by the constitutive expression of a neutralizing gene. Whole genomes of several plant parasitic nematodes were completely sequenced *viz.*, *Meloidogyne incognita, M. hapla, Bursaphelenchus xylophilus* and some of them are in the process of completion *viz., M. javanica, G. rostochiensis, G. pallida, Heterodera schachtii, Radopholus similis etc.* Genomics approach can be used to identify novel gene targets in the nematode genome, which were responsible for maintaining nematode-specific relationships and on how these genes can be targeted to inhibit nematode growth and development. Research accomplishments in genomics and molecular biology include the discovery of heat shock protein genes possibly involved in developmental arrest of the soybean cyst nematode, the identification of neuropeptides and female-specific proteins in the soybean cyst nematode, the disruption of nematode reproduction with inhibitors of nematode sterol metabolism, the development of novel morphological and molecular (heat shock protein genes and the D3 segment of large subunit ribosomal DNA) features useful for nematode identification and classification, and the elucidation of the population genetics of potato cyst nematode pathotypes.

Phytochemical based Approaches

Higher plants have yielded a broad spectrum of active compounds, including polythienyls, isothiocyanates, glucosinolates, cyanogenic glycosides, polyacetylenes, alkaloids, lipids, terpenoids, sesquiterpenoids, diterpenoids, quassinoids, steroids, triterpenoids, simple and complex phenolics, and several other classes. Many other antinematicidal compounds have been isolated from antagonistic organisms. Avermectins produced by *Streptomyces avermetilis*, a soil inhabiting actinomycete, has a great potential and is nematicidal, insecticidal and acaricidal (Putter *et al.*, 1981). Natural products active against mammalian parasites can serve as useful sources of compounds for examination of activity against plant parasites. Phytotoxic approach can also be done by expressing a phytotoxic gene product specifically in the feeding cells formed by plant parasites, so causing either specific cell death or limiting nematode feeding by attenuating the feeding cell. Biofumigation using glucosinolate containing plants as biologically active rotation and green manure crop also control soil borne pathogens and plant parasitic nematodes. Utilization of phytochemicals and phytochemical based approaches, although currently uneconomic in many situations, has a tremendous potential in managing plant parasitic nematodes.

Newer and Novel Molecules, Formulations and Delivery Systems of Chemical based Approaches

Currently research on newer and novel molecules is aimed to identify new gene targets in nematode metabolism besides identifying analogs of the original chemical leads with improved nematicidal properties while maintaining activity and low phytotoxicity. Many chemical plant defense activators rarely found in plants like BABA(DL-β-aminobutyric acid), BTH(benzo(1,2,3) thiodiazole-7-carbothioic acid S-methyl ester) can be used to activate plant defense mechanisms against biotic and abiotic stresses without negative influence on plant growth. Biological plant defense activators, usually the elicitors from antagonistic organisms like lipopolysaccharides, chitin oligomers, glucans, siderophores and enzymes like xylanases can be used as initial signaling molecules to activate plant defense mechanisms to combat biotic stress (Parvatha Reddy, 2013). The availability of thousands of new nucleotide sequences of plant parasitic nematodes in public databases helps to identify and characterize our genes of interest, including those encoding pathogenicity factors, diagnostic markers, targets (for vaccines, drugs, or nematicides), and immunomodulatory molecules (Mc Carter *et al.*, 2005). These gene targets will be used to design gene-specific inhibitors that inhibit important biological functions in the invading nematode or prevent the nematode from organizing and obtaining nutrients from the feeding site. The new target molecules are for ovicidal, moulting inhibition, cuticle and collagen inhibitors, nematode surface binding etc. The mechanism of action of these compounds will be verified both by biochemical and molecular approaches.

RNA-Mediated Interference or Gene Silencing Approaches

RNA interference is a mechanism for RNA-guided regulation of gene expression in which double-stranded ribonucleic acid inhibits the expression of genes with complementary nucleotide sequences (Karakas, 2008). It has most likely evolved as a mechanism for cells to eliminate foreign genes that result in sequence-specific inhibition of gene expression at the transcription, mRNA stability or translational levels (Mello and Conte, 2004). RNAi can be used for systematic shut down of each gene in the cell, which can help identify the components necessary for a particular cellular process or an event such as cell division. The most interesting aspect of RNAi is that, it is highly precise, remarkably powerful and the interfering activity can cause interference in cells and tissues even far from the site of introduction. It was observed that the injection of double-stranded RNA (dsRNA) into *C. elegans* corresponding to any one of its genes specifically depleted the mRNA molecules transcribed from that gene (Fire *et al.*, 1998). In the absence of mRNA, the corresponding protein is not produced and, therefore, this ultimately results in the loss of function of the targeted gene. The cells of organisms such as nematodes, plants and human do not produce dsRNA. However, many viruses produce dsRNA. Thus, the RNAi machinery seems to have evolved as a defense mechanism against viruses. Scientists have exploited RNAi (gene silencing) as a powerful tool to disrupt the functions of specific genes to uncover their functions. Since its original discovery in *C. elegans*, RNAi has been observed in many other organisms as well (Hannon, 2003). In *C. elegans*, RNAi was triggered by means of microinjection, feeding worm with bacteria transformed to produce dsRNA for targeted nematode genes (Timmons *et al.*, 2001) and soaking worms in dsRNA solution. In plant parasites, any of the above mentioned method may be challenging because of obligatory parasitic nature and inconsistency of transformation. Soaking worms in dsRNA solution gave better results than the other two methods mentioned above for plant parasitic nematodes. Lilley *et al.* (2012) used soaking method of delivery for eggs and second stage juveniles of sedentary nematodes species for *in vitro* RNAi. Silencing of target genes through *in vitro* RNAi in PPNs was first achieved in infective juveniles of the cyst nematodes, *Globodera pallida* and *Heterodera glycines*, through neuro stimulant-mediated oral ingestion of dsRNA molecules (Urwin *et al.*, 2002). *In vitro* RNAi strategy successfully down regulated many of targeted genes with distinct phenotypic and biological changes and the results have been demonstrated successfully in root-knot, cyst, lesion, pine wilt, burrowing, and white tip nematodes (Lilley *et al.*, 2012; Cheng *et al.*, 2013; Tan *et al.*, 2013). After the successful application of *in vitro* RNAi in plant parasitic nematodes, research has been started to study host induced gene silencing where the dsRNAs will be produced by the host system for silencing the targeted gene in pathogen system. Yadav *et al.* (2006), successfully demonstrated host induced gene silencing in transformed tobacco plants, where plants produced double stranded RNAs for silencing two essential genes (*splicing factor* and *integrase*) of the parasitic nematode, *M. incognita* and a significant reduction in nematode infection was observed. The transgenic tobacco plants very effectively resisted *M. incognita* infection and their development. These nematodes were specifically deficient in the mRNA of targeted genes, indicating that the dsRNA produced in plants did indeed trigger RNAi response in the nematode.

A similar strategy was deployed to silence the genes encoding a β-subunit of the coatomer (COPI) complex and a pre-mRNA splicing factor, resulting in the reduced infection and development of *H. glycines* on the transgenic soybean (Li *et al.*, 2010 a,b). Pioneer genes of specific nematode are the attractive targets for host induced RNAi. According to Steeves *et al.* (2006), transgenic soybeans expressing siRNAs specific to a major sperm protein (*MSP*) gene of soybean cyst nematode, *H. glycines* suppressed reproduction of the nematode significantly. More recently, tomato hairy roots expressing a hairpin of the *mj-far-1* (fattyacidandretinol binding protein) gene of *M. javanica*, led to 80 per cent reduction in transcript abundance in the feeding nematodes and a significant reduction in giant cell number due to impaired development of females (Iberkleid *et al.*, 2013). When Sindhu *et al.* (2009) studied the effect of *in planta* RNAi against four parasitism genes of cyst nematodes, a significant reduction in parasitism was observed. Ectopically expressed double stranded RNAs targeted for silencing two genes (*Miduox, Mispc3*) from root knot nematode *M. incognita* significantly reduced nematode development in transformed *Arabidiopsis* plants (Charlton *et al.*, 2010). Soybean plants expressing an RNA hairpin targeting the tyrosine phosphatase gene of *M. incognita* supported 92 per cent fewer galls than control plants (Ibrahim *et al.*, 2011). When the *M. incognita* calreticulin gene, *Mi-CRT* (Jaouannet *et al.*, 2013) and a parasitism gene, *8D05* (Xue *et al.*, 2013), were targeted for RNAi in Arabidopsis plants, a significant reduction in gall number and developing females were observed.

The present technology allows the transformation of an increasing number of crop plants, providing new ways to introduce resistance against plant-parasitic nematodes. The ability of sedentary nematodes to induce specialized feeding sites in plant roots is one of the most fascinating aspects of this host-parasite interaction. Molecular approaches have been developed to identify and characterize plant genes altered in expression after infection by sedentary nematodes. The results obtained indicate that many genes indeed become up-regulated upon nematode infection. Surprisingly, several so-called constitutive promoters that are normally used to achieve high expression in plant cells are completely 'silenced' in the feeding sites within days after nematode infection. A profile of 37 distinct putative parasitic genes have been identified by Huang *et al.* (2003) expressed exclusively within the esophageal gland cells of sedentary endoparasite, *M. incognita*. With the advent of RNAi technology it is now possible to silence a gene throughout an organism or in specific tissues. This offers the versatility to partially silence or completely turn off genes, work in both cultured cells and whole organisms and can selectively silence genes at particular stages of the organism's life cycle (Shahinul Islam *et al.*, 2005). In India, ICAR-Indian Agricultural Research Institute is conducting research on management of root knot nematodes using RNAi. Parasitism genes of nematode esophageal glands are being targeted and functional validation of different genes like *flp-14*, *flp-18* (FMRFamide-like peptide genes) (Papolu *et al.*, 2013), *16D10* (subventral esophageal gland specific gene) were done using RNA interference. Now the complete DNA sequence (genome) is available for many plant parasitic nematodes where RNAi can be applied for revealing the biological functional information contained in their genomes. In fact, this has already revolutionized the field of functional genomics, an area of biology focused on the characterization of gene function in large scale.

Horticultural Practices and Approaches for Nematode-Free Planting Material

It is widely proven that clean and healthy planting material (short-duration transplants, rooted perennials, cuttings, tissue cultured plantlets etc.) have direct influence on initial plant potential either in open fields or in protected hi-tech polyhouses, which decides the extent of realization of genetic potential of the crop in terms of production and productivity. In many of the polyhouse raised tomato, English cucumber, bell pepper, carnation etc., root-knot nematodes devastated leading to abandoning of the polyhouse facilities/structures. To deal with this, horticultural practice of grafting the desired variety on to resistant root-stock grafted plantlets is widely being adopted and mechanized nursery units with grafting facilities have become popular in East-Asian countries especially in bell pepper, carnation and English cucumber. Since grafting of high yielding susceptible perennial on to resistant root-stock is a highly successful method for the management of nematode problems in perennials. Similar grafting technologies need to be replicated on a commercial scale to address root-knot nematode menace in pomegranate, citrus and grapevine on a regular basis. Biofumigation using plants belonging to the families Brassicaceae (cabbage, cauliflower, kale and mustard), Capparidaceae (Cleome) and Moringaceae (horse radish) is a very efficient method to manage nematodes in horticultural crops. Biofumigation using these plants works with an efficient and important enzymatic defensive system called Myrosinase-glucosinolate system (Parvatha Reddy, 2013). Crop rotation with non host crops or antagonistic crops and soil solarization are some of the specific practices that can regulate the soil populations of parasitic nematodes and soil-borne pathogens in horticultural systems.

Evolving and Wider Adoption of Good Agricultural Practices (GAP) Approach

There is a need to evolve or formulate Good Agricultural Practices for each crop situation, which can in general ward off or minimize losses due to biotic stresses. GAP in combination of integrated pest management with biological means, minimized use of target specific chemicals and use of only registered chemicals, using resistant cultivars can assure sustained plant and soil health for better realization of crop productivity. Hence, an integrated management strategy will be the most effective in managing the nematode problems in a safe and ecologically acceptable manner.

References

Al-Raddad A M 1995. Interaction of *Glomus mosseae* and *Paecilomyces lilacinus* on *Meloidogyne javanica* of tomato. *Mycorrhiza* **5:** 233-236.

Ashoub A H and Amara M T 2010. Biocontrol Activity of Some Bacterial Genera Against Root-Knot nematode, *Meloidogyne incognita*. *Journal of American Science* **6:** 321-328.

Charlton W L, Harel H Y M, Bakhetia, Hibbard J K, Atkinson H J and Pherson M J 2010. Additive effects of plant expressed double-stranded RNAs on root-knot nematode development. *International Journal of Parasitology* **40:** 855-864.

Chen R, Li H, Zhang J, Xiao J and Zhibiao Y 2007. *CaMi*, a root knot nematode resistance from hot peer *Capsicum annum* L. confers nematode resistance in tomato. *Plant Cell Reporter* **26:** 895-905.

Cheng X, Xiang Y, Xie H, Xu C L, Xie T F and Zhang C 2013. Molecular characterization and functions of fatty acid and retinoid binding protein gene (*Ab-far-1*) in *Aphelenchoides besseyi*. *PLoSONE* **8:** 66011.

Claverie M, Dirlewanger E, Bosselut N, Ghelder C, Voisin R, Kleinhentz M, Lafargue, B, Abad P, Rosso M N, Chalhoub B and Esmenjaud D 2011. The *Ma* Gene for Complete-Spectrum Resistance to *Meloidogyne* species in Prunus is a TNL with a Huge Repeated C-Terminal Post-LRR Region. *Plant Physiology* **156:** 779-792.

Elling A A 2013. Major emerging problems with minor *Meloidogyne* species. *Phytopathology* **103:** 1092-1102.

Felix M A, Ashe A, Piffarretti J, Wu G, Neuz I, Belicard T, Jiang Y Zhao G, Franz C J, Goldstein L D, Sanroman M, Miska E A and Wang D 2011. Natural and experimental infection of *Caenorhabditus* nematodes by novel viruses related to noda viruses. *PLoS Biology* 9: 1-14.

Fire A, Xu S, Montgomery M K, Kostas S A, S E Driver and Mello C C 1998. Potent and specific genetic interference by double-stranded RNA in *Caenorhabditis elegans. Nature* 391: 806-811.

Goswami B K, Pandey R K, Rathour K S, Bhattacharya C and Singh L 2006. Integrated application of some compatible bioagents along with mustard oil seed cake and furadan on *Meloidogyne incognita* infecting tomato plants. *Journal of Zhejiang University Science* B 7: 283-285.

Hannon G J 2003. *RNAi: A guide to gene silencing,* Cold Spring Harbor: Cold Spring Harbor Laboratory Press.

Huang G, Gao B, Maier T, Allen R, Davis E L, Baum T J and Hussey R A 2003. A profile of putative parasitism genes expressed in the esophageal gland cells of the root-knot nematode *Meloidogyne incognita. Molecular Plant-Microbe Interaction* 16: 376-381.

Hunter-Fujita F R, Entwistle P F, Evans H F and Crook N E 1998. Insect viruses and pest management. John Wiley and Sons, Chichester, New York.

Hussey R S and Janssen GJW 2002. Root-knot nematodes: *Meloidogyne* species. In: Starr, *et al.* (ed.) Plant resistance to parasitic nematodes. CAB International Press, United Kingdom. pp. 43-70.

Iberkleid I, Vieira P, de Almeida Engler J, Firester K, Spiegel Y and Horowitz S B 2013. Fatty Acid and Retinol-Binding Protein, *Mj-FAR-1* induces tomato host susceptibility to root-knot nematodes. *PLoS ONE* 8: e64586

Ibrahim H M, Alkharouf N W, Meyer S L, Aly M A and Gamal E A K 2011. Post-transcriptional gene silencing of root-knot nematode in transformed soybean roots. *Experimental Parasitology* 127: 90-99.

Jaouannet M, Magliano M, Arguel M J, Gourgues M and Evangelisti E 2013. The root-knot nematode calreticulinMi-CRT is a key effector in plant defense suppression. *Molecular Plant Microbe Interactions* 26: 97-105.

Karakas M 2008. RNA interference in plant parasitic nematodes. *African Journal of Biotechnology* 7: 2530-2534.

Kerry B R 1984. Nematophagous fungi and the regulation of nematode populations in soil. *Helminthological Abstract Series* B 53: 1-14.

Khan M R, Jain RK, Singh RV and Pramanik A 2010. In Economically important plant parasitic nematodes atlas. p. 2.

Li J, Todd T C, Oakley T R, Lee J and Trick H N 2010a. Rapid inplanta evaluation of root expressed transgenes in chimeric soybean plants. *Plant Cell Reporter* 29: 113-123.

Li J, Todd T C, Oakley T R, Lee J and Trick H N 2010b. Host-derived suppression of nematode reproductive and fitness genes decreases fecundity of *Heterodera glycines* Ichinohe. *Planta* **232**: 775–785.

Lilley C J, Davies L J and Urwin P E 2012. RNA interference in plant parasitic nematodes: a summary of the current status. *Parasitology* **139**: 630-640.

McCarter J P, Mc K Bird D and Mitreva M 2005. Nematode gene sequences: Update for December 2005. *Journal of Nematology* **37**: 417-421.

Mello CC and Conte D J 2004. Revealing the world of RNA interference. *Nature* **431**: 338-342.

Milligan S B, Bodeau J, Yanghoobi J, Kaloshian I, Zabel P and Wiiliamson V M 1998. The root knot nematode resistance gene *Mi* from tomato is a member of the leucinezppier, nucleotide binding, luecine rich repeat family of plant genes. *Plant Cell* **10**: 1307-1319.

Papolu P K, Gantasala N P, Kamaraju D, Banakar P, Sreevathsa R and Rao U 2013. Utility of host Host Delivered RNAi of Two FMRFamide like peptides, filp-14 and flp-18, for the management of root knot nematode, *Meloidogyne incognita*. *PLoS ONE* **8**: e80603.

Parvatha Reddy P 2013. Recent advances in crop protection. Ist ed. Springer, India.

Putter J G, Mac Connell F A, Preiser F A, Haidri A A, Rishich S S and Dybas R A 1981. Avermectins: novel class of insecticides, acaricides and nematicides from a soil microorganism. *Experientia* **37**: 963-964.

Rao M S, Umamaheswari R and Chakravarty AK 2015. Plant Parasitic Nematodes: a major stumbling block for successful crop protection under protected conditions in India. *Current Science* **108**: 13-14.

Roberts P A 2002. Concepts and consequences of resistance. In Starr *et al.* (ed.) Plant resistance to parasitic nematodes. CAB International Press, Wallingford, UK. pp23-41.

Sasser J N and Freckman D W 1987. A world perspective on Nematology: The role of the society. In Vistas on Nematology(Eds J A Veech and D W Dickson). Society of Nemtologists, Hyattsville, Maryland pp7-14.

Satyandra S and Nita M 2010. *In vitro* studies of antagonistic fungi against the Root-knot nematode, *Meloidogyne incognita*. *Biocontrol Science and Technology* **20**: 3-4.

Shahinul Islam S M, Miyazaki F T and Itoh K 2005. Dissection of gene function by RNA silencing. *Plant Biotechnology* **22**: 443-446.

Sindhu A, Maier T R, Mittchum M G, Hussey R S, Davis, L E and Baum J T 2009. Effective and specific *in planta*RNAi in cyst nematodes: Expression interference of four parasitism genes reduces parasitic success. *Journal of Experimental Botany* **1**: 315-324.

Steeves R M, Todd T C, Essig J S and Trick H N 2006. Transgenic soybeans expressing siRNAs specific to a major sperm protein gene suppress *Heterodera glycines* reproduction. *Functional Plant Biology* **33**: 991-999.

Stirling G R 2014. Biological control of Plant Parasitic Nematodes-soil ecosystem management in sustainable agriculture. 2nd ed.UK,Oxfordshire,UK: CAB International press.

Tan C H J, Jones M G K and Fosu N 2013. Gene silencing in root lesion nematodes (*Pratylenchus* spp.) significantly reduces reproduction in a plant host. *Experimental Parasitology* **133:** 166-178.

Timmons L, Court D L and Fire A 2001. Ingestion of bacterially expressed dsRNAs can produce specific and potent genetic interference in *Caenorhabditis elegans*. *Gene* **263:** 103-112.

Tribe H R 1977. Pathology of cyst nematodes. *Biological Reviews* **52:** 477-507

Urwin P E, Lilley C J and Atkinson H J 2002. Ingestion of double-stranded RNA by preparasitic juvenile cyst nematodes leads to RNA interference. *Molecular Plant Microbe Interactions* **15:** 747-752.

Wiliamson VM 1998. Root knot resistance genes in tomato and their future use. *Annual Review of Phytopathology* **36:** 277-293.

Williamson V M and Kumar A 2006. Nematode resistance in plants: The battle underground. *Trends in Genetics* **22:** 396-403.

Xue B, Hamamouch N, Li C, Huang G and Hussey R S 2013. The 8D05 parasitism gene of *Meloidogyne incognita* is required for successful infection of host roots. *Phytopathology* **103:** 175-181.

Yadav B C, Veluthambi K and Subramaniam K 2006. Host generated double stranded RNA induces RNAi in plant- parasitic nematodes and protects the host from infection. *Molecular and Biochemical Parasitology* **148:** 219-220.

Chapter 2

Behaviour Modifying Chemicals and their Utility in Pest Management

A R Prasad, A L Prasuna and K N Jyothi

*Centre for Semiochemicals, CSIR-Indian Institute of Chemical Technology (IICT),
Hyderabad – 500 007, Telangana State
E-mail: arp@iict.res.in*

ABSTRACT

Behaviour modifying chemicals, regularly referred as semiochemicals, are the chemical signals that transmit information among the individuals. Attraction of insects to plants and other host organisms involves detection of specific semiochemicals in specific ratios. This natural instinct in insects towards the semi chemicals is utilized by entomologists and chemists to formulate an environmentally safe and eco-friendly pest management strategy. Semiochemical based pest management practices are now more acceptable not only over synthetic conventional pesticides but also over botanically-derived toxicants including the genetically modified organisms (GMO). These chemicals provide pest control strategies by regulating the behaviour and development of the pest insect rather than toxicity or any direct physiological effects. Semiochemicals are divided into two major groups of chemicals, pheromones that mediate intraspecific interactions and allelochemicals, which mediate interspecific interactions. At present the most widely used semiochemicals in pest management are insect sex pheromones particularly those of lepidopteran pests. Pheromones are naturally produced organic chemicals released from the organisms in minute quantities elicit behavioural responses from the member of same species. In insects, these pheromones are used in many behavioural contexts, e.g. attraction in conspecifics for mating (sex pheromones), congregation for food or reproduction (aggregation pheromones), alarm pheromones that serve insects on attack of predators, deterrence towards oviposition sites (epideictic pheromones) and trial marking pheromones (trial pheromones). Since the discovery of first pheromone from Bombyx mori in 1959, hundreds of pheromones are identified till date and great research has been carried out on synthetic sex pheromones of the order Lepidoptera for their practical utility. Pheromones are now well utilized as monitoring tools in integrated pest management (IPM) programmes. Pheromones are also useful as means of pest population reduction through mass trapping and mating disruption. Isolation, identification, synthesis of insect pheromones and formulating them for field use to control insects themselves is termed as Pheromone Application Technology (PAT). This kind of approach needs an interdisciplinary collaboration between entomologists, chemists, technologists and finally farmers.

The Centre for Semiochemiclas at Indian Institute of Chemical Technology (IICT), perhaps the only centre in India with its multidisciplinary expertise is actively engaged in all aspects of pheromones research and developed PAT for several important insect pests of field crops. IICT, in collaboration with agriculture based institutions has successfully demonstrated the field efficacy of indigenously developed PAT for the management of important crop pests. The present review focuses on successful efforts made by IICT in developing and implementing PAT for the control of groundnut leaf miner (GLM), Aproaerema modicella Deventer (Gelechiidae: Lepidoptera), an important oilseed pest.

Keywords: *Behaviour modifying chemicals/semiochemicals, Pheromones, Pest management.*

Introduction

Pest management with less environmental ill effects and sustainable production of safe, high quality and globally competitive agro products are the main objectives of the modern agriculture. In view of adverse effects manifested due to the indiscriminate use of conventional pesticides ecologically benign pest management strategies that contribute to Integrated Pest Management (IPM) are given priority worldwide (Meerman, 1996). Development and utilization of semiochemical based technologies for pest management has shown a path to reach a sustainable eco-friendly and environmentally safe agro practice (Mitchell, 1986).

Semio chemicals are chemical signals that are produced by plants or animals for specific communication purposes with other individuals. Attraction of insects to plants and other host organisms involves detection of specific semiochemicals (pheromones and kairomones), the natural signal chemicals often referred as behaviour modifying chemicals in specific ratios. Pheromones are organic chemicals released by the organisms in minute quantities for communication purpose within their own species. In insects, these pheromones are used in many behavioural contexts, *e.g.* attraction in conspecifics for mating (**sex pheromones**), congregation for food or reproduction (**aggregation pheromones**), dispersal on attack of predators (**alarm pheromones**), deterrence towards oviposition sites (**epideictic pheromones**) and **trial** marking **pheromones** (Jutsum and Gordon, 1989). Since semiochemicals act as signals to regulate/modify the behaviour of the receiving individual they are also referred as behaviour modifying chemicals and the management strategies mostly exploit the resulted behaviour manipulations. Attempts for exploring the possibilities for exploitation of semiochemicals/behaviour modifying chemicals as pest control agents proved them as invaluable tools in modern IPM programmes (Beroza, 1972). Working more on the behaviour modifying nature of semiochemicals, researchers recognized that the long-range efficacy of sex pheromones could be exploited for pest management.

Insect pheromones that have been thoroughly studied and most successfully exploited in pest monitoring/control are lepidopteran **sex pheromones** followed by coleopteran **aggregation pheromones** (Arn *et al.,* 2000). Sex pheromones are chemical scents secreted either by female or male (usually by female) insects and attract the opposite sex for mating. Sex pheromones appear to be distributed ubiquitously throughout the class of insecta and have been identified from at least 10 orders (Metcalf, 1998). The potential of sex pheromones particularly of insects that belong to Lepidoptera, a very important economic insect order, has been well exploited globally in pest management practices (Rosca *et al.*, 1985). Sex pheromone based technology is now internationally recognized as green plant protection technology marked with ecological security and sustainability. The present article focuses on the major features of insect sex pheromones, their modes of application as pest control agents while centring the successful efforts of IICT in developing and implementing PAT for the control of groundnut leaf miner (GLM), *Aproaerema modicella* Deventer (Gelechiidae: Lepidoptera) an important pest of groundnut.

Pheromone Application Technology (PAT)

The corner stone of pheromone based pest control is its well defined/identified pheromone system. In moths, multiple constituents present in a fixed ratio mostly form this system. Practical application of pheromones usually requires specific active chemicals to be isolated, identified and produced/synthesized. The first such pheromone identification was that of 'Bombykol' from *Bombyx mori* (L.), the silk moth, by Butenandt *et al.* (1959). However, our understanding about semiochemicals has been accelerated in the last two decades only after the advancement of technical aspects in various disciplines. Analytical chemistry and Electrophysiology, which utilize Electroantennography (EAG) and EAG coupled Gas chromatography (GC-EAD), are the best examples amongst them. These unique techniques made possible to determine the physiologically active compounds from a complex odour mix (Arn *et al.*, 1975, Hummel and Miller, 1984). Insect antenna or the whole insect can be utilized as an olfactory biosensor to detect and locate the odour of interest.

Developing the synthetic mimics for the identified natural pheromone chemicals through organic synthesis and formulating them for the field use to the control of the same insect pest is **Pheromone Application Technology** (PAT). This technology is associated with species specificity, safety to non-target organisms, ease of use and efficacy. In fact, PAT is a powerful tool in Integrated pest management (IPM) strategies and brings about a long term reduction in pest populations that cannot be accomplished with conventional insecticides. Especially PAT is found highly suitable and the one and only ideal tool for control of insect pests like **leaf miners, webbers, fruit borers, dwellers** and **storage pests**. Insect pheromones, with their unique mode of action, provide appropriate and effective management for these pests targeting the adult insects. Pheromone application is best suitable in organic farming as they are complementary to other control strategies. Pheromones in IPM programmes proved mainly to rationalize the use of conventional insecticides with the reductions of up to 50 per cent (Baker *et al.*, 2002).

Chemistry of Pheromones

Pheromone is a Latin (Greek) word. *Phereum* means to carry and *horman* means to excite or stimulate (Karlson and Fuscher, 1959). More or less all the insect pheromones are volatile compounds. Over the last 40 years, scientists have identified pheromones from over 1,500 different species of insects (Arn *et al.*, 2000). Structurally, the sex pheromones of Lepidoptera are multi component aliphatic linear unsaturated compounds from 10 to 20 carbon atoms being the most frequent 12, 14 and 16 with alcohol, ester, aldehyde, ketone or epoxide functionality at the end of carbon chain. They have one or more double bonds at various carbon numbers. These are probably derived biosynthetically from their respective fatty acids. Other types of pheromones may have more complex structures, often terpenoids but generally contain only carbon, hydrogen and oxygen. Insects do achieve species specificity for the pheromones through the following:

☆ Basic chemical structure,

☆ Position and configuration of double bonds and Optical isomerism (Chirality, where often only one optical isomer is active) and Defined blend ratio of Pheromone components.

Common Structures of Pheromones

(a) Unsaturated Straight Chain Aliphatic Alcohols, Acetates, Aldehydes, Lactones etc. (Lepidoptera)

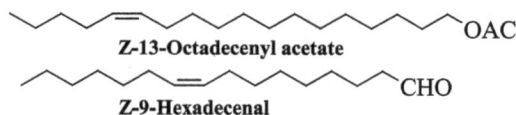

Z-13-Octadecenyl acetate

Z-9-Hexadecenal

(b) Intramolecular-linked ketal (Beetles)

(-)frontalin

Exobrevicomin

(c) Spiroketals (wasp, bee, olive fly)

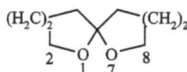

1,7-Dioxa-spiro[5.5]undecane

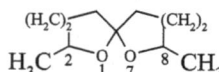

2,8-Dimethyl-1,7-dioxa-spiro[5.5]undecane

(d) Oxygen & Nitrogen Heterocycles (Fine Ants)

3-Heptyl-5,7a-dimethyl-hexahydro-pyrrolizine

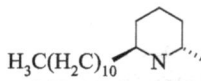

2-Methyl-6-undecyl-piperidine

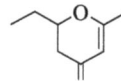

2-Ethyl-6-methyl-2,3-dihydro-pyran-4-one
Hepialone

e) Terpenoids (Bollweevils, anthropods, termites etc.)

Grandisol, Ipsenol, several sesqui and diterpenes

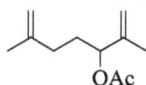

OAc
Acetic acid 1-isopropenyl-4-methyl-pent-4-enyl

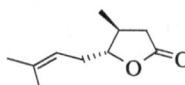

4-Methyl-5-(3-methyl-but-2-enyl)-dihydro-furan-2-one

In general, Lepidopteron insects, specially the insects belong to Noctuidae, do possess a very strong sex pheromone communication system to find their mate for their offspring development. The mate finding sex pheromone components in Lepidoptera are very simple straight chain alcohols, acetates and aldehydes. Lepidoptera sex pheromone components have been identified from more than 500 species and sex attractants are reported for approximately 1500 species (Arn *et al.*, 2000).

Pheromones are a primary method of communication in insects, by which the chemicals serve to identify members of the same species and thus possible mates. The sex pheromone, which is secreted by the adult (generally by a female moth) for the benefit of a specific partner, plays an important role in reproductive isolation. In fact, the "calling" female produces the following pattern of response in a resting male moth through the release of pheromone blend:

☆ Reception-antennal elevation or twitching

☆ Activation-wing fanning or fluttering

☆ Active flight

☆ Orientation to the source of pheromone *i.e.* anemotaxis

☆ Alighting-landing in the immediate vicinity of the female moth

☆ Courtship-including gland extraction and release of male pheromone if any?

The synthetic chemicals should mimic the pheromone blend of the given insect and must produce the above complexial sequence of response in male insects. Over the last 40 years, chemists have extracted and analyzed natural insect pheromones from over 1,500 different insect species and developed synthetic routes to prepare most of these chemicals in large quantities to use them in the fields.

Techniques of Pheromone Application

Pheromones have found application in three major ways known as **monitoring, mass trapping and mating disruption techniques.**

Monitoring

Pest monitoring provides data about the presence and density of the target pest species. Monitoring helps in determining when and how often to time control actions. Pest monitoring minimizes environmental pollution and the costs for control of pests by avoiding the unnecessary pesticide sprayings. In the monitoring approach a small number of traps (3 to 5 per ha) loaded with sex pheromone of a specific pest are placed in the field to attract the opposite sex. The traps act as specific and sensitive indicators regarding the targeted pest. Lepidopteran pest populations in agricultural systems are monitored by collecting males in traps baited with synthetically – produced female pheromones (Roelofs and Carde, 1977).

Mass Trapping

This technique is used to capture and trap out the large number of insects. It is basically the same as monitoring. The difference lies in the number of traps placed in the field (10 to 20 per ha). Success is depended on the number of traps installed, ecology and life cycle of the insect, reproductive behaviour of the pest and the type of the pheromone used in the trap Mass trapping technique is particularly successful against coleoptera wherein the aggregation types of pheromones are employed for trapping both sexes of target pests. Aggregation pheromone is similar to sex pheromone but attractive to both sexes and thus help in locating food sources, hibernation sites, oviposition sites and also to some extent mating as it causes con specifics to gather at one site (Ali and Morgan, 1990). Aggregation pheromones offer excellent potential for direct control in pest management when incorporated into traps along with host plant volatiles and currently being used to control forest pests (Blight *et al.*, 1984; Dickerson *et al.*, 1987).

Mating Disruption

Mating disruption is used typically with female sex pheromone of Lepidoptera. The technique involves the permeation of the air around the crop with large quantity of pheromone to prevent the male moth to locate the female successfully for mating. The exact mechanism underlying the technique is still not clear, but one of the following possible explanations is thought to be involved (Carde and Minks, 1995):

- ☆ Sensory fatigue, representing either adaptation or habituation;
- ☆ Competition between the pheromone dispensers and the calling insect (false trail)
- ☆ Masking of the natural pheromone concentration.

Currently, 30 mating disruption pheromone based products are registered by the US-EPA for the control of eleven lepidopterous pest species like pink bollworm, codling moth, oriental fruit moth, gypsy moth, grape berry moth *etc.*, Commercial slow release formulations were made available for sex pheromones of pink bollworm for controlling the pest by mating disruption in early 1980's (Critchley *et al., 1985*).

Mass trapping and mating disruption both interfere with the mating system of insect pests and reduce pest populations in the subsequent generations and are considered as direct control techniques (Carde and Elkinton, 1984; Carde and Minks, 1995; Cork and Hall, 1998). Currently, > 30 mating disruption pheromone based products are registered by the US-EPA for the control of eleven lepidopterous pest species like pink bollworm, codling moth, oriental fruit moth, gypsy moth, grape berry moth *etc.* All over the world, the crop protection councils are encouraging the eco-friendly application of insect pheromones through IPM as one of the best alternatives for conventional insecticides. However, utility of pheromones in pest management has remained as a new concept in the Indian farming and is restricted to monitoring trials only particular to few insect pests. Centre for semiochemicals at Indian Institute of Chemical Technology (IICT) with its multi-disciplinary expertise had taken a lead role to bring this eco friendly technology into practice and popularization. In this direction, IICT developed pheromone mediated control strategies up to mass trapping for several insect pests of important crops like rice, sugarcane, brinjal, groundnut etc. The following describes the efforts made by IICT in developing PAT for GLM control as an alternative to pesticides.

Pheromone Application Technology (PAT) for Groundnut Leaf Miner (GLM)

Aproaerema modicella (Deventer) (Gelechiidae: Lepidoptera), the groundnut leaf miner (GLM) is a serious pest on groundnut in South and South East Asia and has been studied most intensively in India (Shanower *et al.*, 1993). Nearly 60 per cent yield loss is reported due to GLM infestation. The larvae feed on foliage mining into the leaf lets and subsequently seize the plant growth. Conventional chemicals found less effective against this internal feeder. Hence, alternative pest control methods like pheromone application targeting adult moths are sought for the management of GLM. Sex pheromone of GLM constitutes three components: Z-7, 9-decadienylacetate (I) E–7 decenyl acetate (II) and Z-7-decenyl acetate (III) in 100:20:14 ratios (Hall *et al.*, 1994). Synthetic methodologies for the production of pheromone components in large quantities up to 50g batch size were developed at IICT (Yadav *et al.*, 1995).

Structures of GLM Pheromone Components

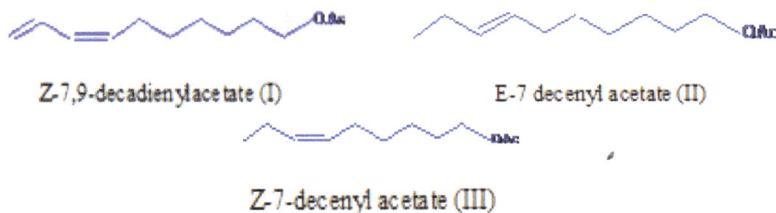

Z-7,9-decadienylacetate (I)   E-7 decenyl acetate (II)

Z-7-decenyl acetate (III)

Schemes Developed at IICT for the Synthesis of GLM Pheromone Components.

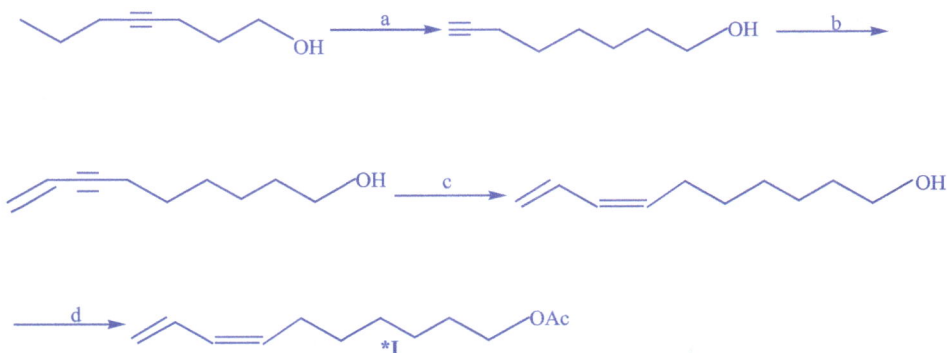

(a) NaNH$_2$, 1,3-Diaminopropane, (b) Diethylamine, CuI, CH$_2$=CHBr, Pd(PPH$_3$)$_4$ (c) Zn-Cu(II) Acetate / AgNO$_3$ (d) Ac$_2$O, DMAP / Pyridine

Synthesis of Z-7, 9- decadienylacetate: Scheme - I

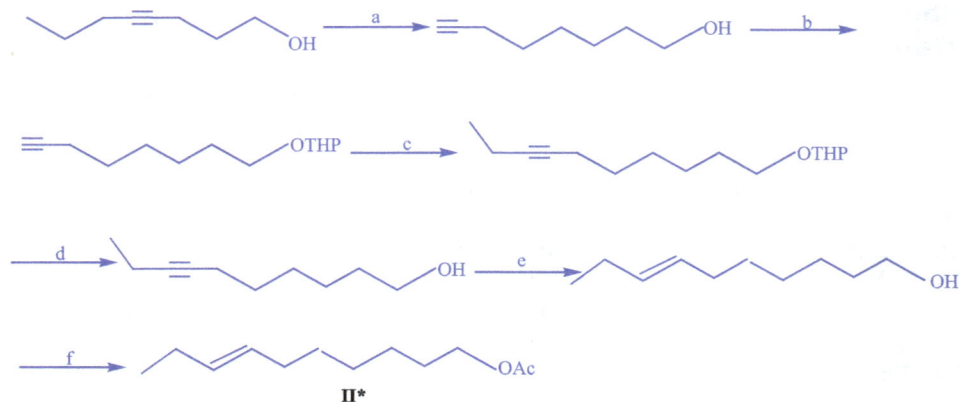

II*

(a) NaNH$_2$/ 1,3-Diaminopropane (b) DHP, PTSA, DCM, (c) Li /Liq NH$_3$, HMPA, EtBr, (d) PTSA, MeOH, (e) Na / Liq NH$_3$ (f) Ac$_2$O, Pyridine/ DMAP

Synthesis of *E*-7-Decen-1-acetate: Scheme- II

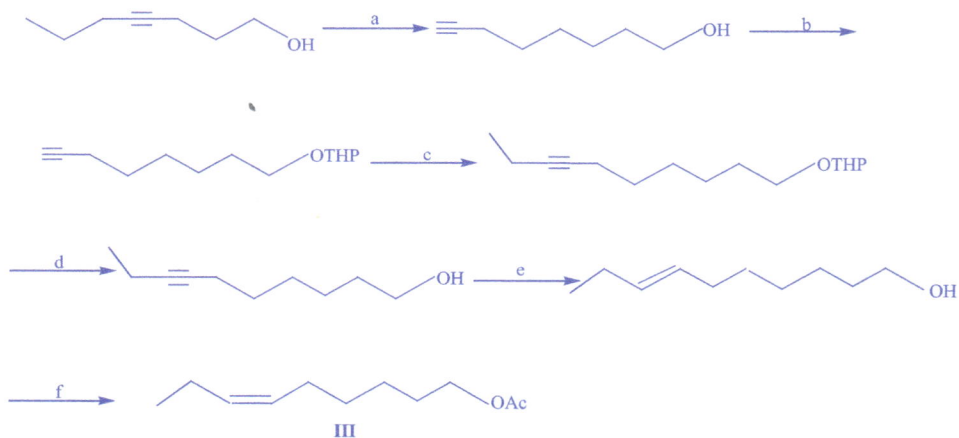

III

(a) NaNH$_2$/ 1,3-Diaminopropane (b) DHP, PTSA, DCM, (c) Li /Liq NH$_3$, HMPA, EtBr, (d) PTSA, MeOH, (e) Pd-CaCO$_3$, H$_2$, EtOH (f) Ac$_2$O, Pyridine/DMAP

Synthesis of *Z*-7 Decenylacetate: Scheme – III

Field Evaluation of Indigenously Developed Synthetic Pheromone

The use of synthetic sex pheromone blend in monitoring GLM populations has been successfully demonstrated in several large scale field trials with farmers' participation during the past 10 years in major groundnut growing districts of A.P. (Yadav *et al.*, 2007). To conduct large scale field demonstrations IICT collaborated with local agricultural based institutions (NRCS, Junagadh; IPE, Hyderabad), Universities (ANGRAU, TNAU etc) and voluntary organizations (KVK: Nalgonda,

SDDPA: Mahboobnagar- Figure 2.1). Village level field demonstrations conducted employing monitoring and mass trapping techniques during last ten years at different locations and the number of insects trapped/trap/week with indigenously developed pheromone technology are as follows:

Table 2.1: Total Acreage Covered at different Locations for Mass Trapping of GLM

Year	Place	Acreage (Ha)	Highest No. Males/Trap/ Std Week
1999-2001	Tindivanam, Tamilnadu	50	<180/35 th –3 8 th std week
2001 and 2002	Gaddipally, Nalgonda Dt.	100	<120/37th to 40 th std week
	Wanaparthy, Mahboobnagar Dt	100	<425/37th to 40 th std week
2006 and 2007	Jalmalkunta, Nalgonda	100200	<350/36th to 40 th std week
	Tellaralla palli, Mahboobnagar		<415/40 th std week
2007-2008	Mahaboobnagar	25	<338/31st to 34th std week
2008-2009	Anantapur	100	367males/trap/51st week
2009-2010	Anantapur	200	<470 males/trap/47th week
2010-2011	Anantapur	300	<253 males/trap/45th week

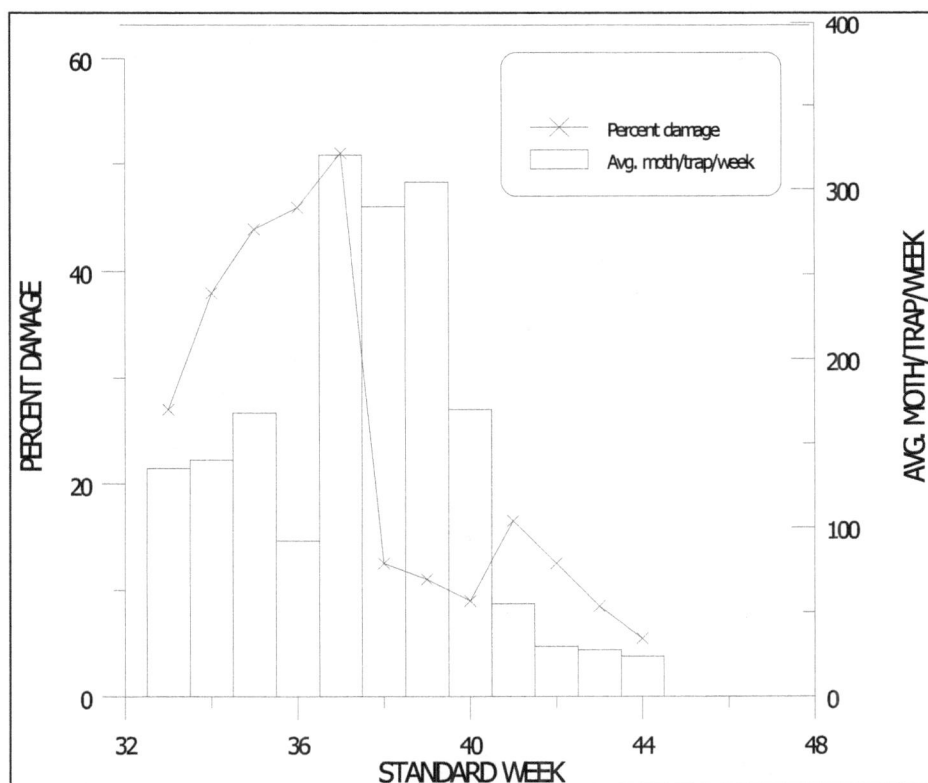

Figure 2.1: Pattern of GLM Catches and the Level of Infestation on Field Application of Pheromone in One Cropping Season Conducted in Mahboobnagar District during 2002.

Commercially available delta sticky trap (modified with replaceable sticky liner for reusing) baited with plastic vial dispenser loaded with 3mg of synthetic sex pheromone blend provided best monitoring for GLM populations up to three weeks.

Mass Trapping

Encouraged with the results we further evaluated the feasibility of pheromone application technology (PAT) for control of GLM through mass trapping in Ananthapur in three continuous years. The pattern of trap catch in monitoring trials during 2007 is similar to our previous monitoring trials in Mahboobnagar and Nalgonda districts conducted during 2003-2004 (Yadav *et al.*, 2007). Results from mass trapping study showed a significant reduction in the total number of moths captured season long from 2007 to 2010 (Figure 2.2). The percent crop damage was also significantly reduced by the end of third year in comparison with control fields. The modified cost effective delta trap with removable sticky liner found to be very useful during the peak flight seasons of moth recording the maximum trap catches. The potential of mass trapping has been successfully demonstrated with significant increase in crop yields/hectare/year in the demonstration fields. Our results proved that the synthetic sex pheromone blend of GLM is a promising tool for suppressing the pest populations through mass trapping technique. Although mass trapping as a control method is seldom used in India, the large number of trapped moths in the present study made participating farmers feel secure and confident about PAT as a reliable tool in pest management. This is the first study that analyzes the potential use of mass trapping with sexual pheromones as a direct control method for GLM in India.

Conclusions

Our study on mass trapping of GLM in Anantapur district is highly successful and the implementing farmers reported that due to implementation of PAT they could save money by not spraying insecticides. They also reported nearly 20 per cent yield increase in PAT treated plots in comparison with control plots. Nearly 1500 farmers from 30 villages were benefited during this programme.

The following recommendations were derived for successful implementation of PAT as a package for GLM control.

☆ GLM Pheromone blend:	(Z) 7, 9 deca dienyl acetate (100); (E) 7 decenyl acetate (20) (Z 7) decenyl acetate (14)
☆ Mode of control:	Monitoring and mass trapping
☆ Number of traps/Ha:	5-7 (monitoring); 15-20 (mass trapping)
☆ Dispenser type:	Plastic vial with lid
☆ Type of trap:	Sticky trap
☆ Pheromone loading:	3 mg/dispenser
☆ Lure replacement:	3 weeks

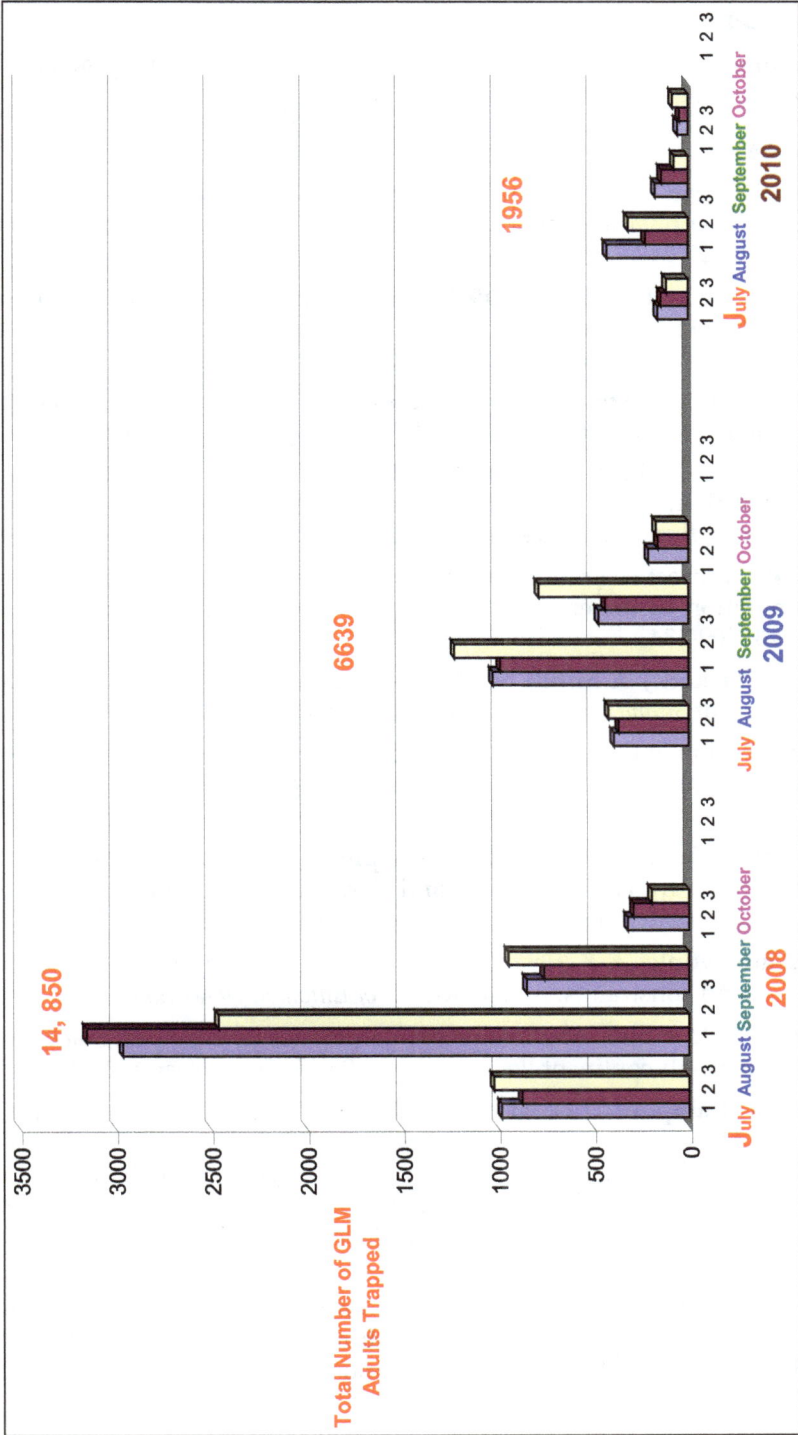

Figure 2.2: Impact of Continuous Pheromone Mass Trapping Technique in Decline of GLM Population at Three Different Villages in Ananthapur District, A.P.

Acknowledgements

The author would like to thank the Director, IICT for providing facilities and also thankful to CSIR-India for providing funding.

References

Ali M F and Morgan E D 1990. Chemical communication in insect communities: a guide to insect pheromones with special emphasis on social insects. *Biological Reviews* **65:** 227-247.

Arn H, Staedler E and Rauscher S 1975. The antennographic detector, a selective and sensitive tool in the gas chromatographic analysis of insect pheromones. *Zeitschrift fur Naturfortschung* **30:** 722-725.

Arn H, Toth M and Priesner E 2000. List of sex pheromones of female Lepidoptera and related male attractants. *Internet Edition:* http: //www-pherolist. slu. se.

Baker B, Benbrook C M, Groth E and Benbrook K L 2002. Pesticide residues in conventional, integrated pest management (IPM) grown and organic foods: insights from three US data sets. *Food Additives and Contaminants* **19:** 427-446.

Beroza M 1972. Insect sex attractant pheromones, a tool for reducing insectcide contamination in the environment. *Toxicology and Environmental Chemical Reviews* **1:** 109-134.

Blight M M, Pickett J A Smith M C and Wadhams L J 1984. An aggregation pheromone of *Sitona lineatus*. *Naturwissenschaften* **71:** 480-481.

Butenandt A Beckmann R Stamm D and Hecker E 1959. Uber den Sexual- Lockstoff des Seidenspinners *Bombyx mori*. Reindarstellung und Konstitution. *Zeitschrift fur Naturfortschung* **14:** 283-284.

Carde R T and Elkinton J S 1984. Field trapping with attractants: Methods and interpretation. In: H. E. Hummel and T A Miller (Editors), Techniques in pheromone research (pp. 111-129). Springer New York.

Carde R T and Minks A K 1995. Control of moth pests by mating disruption: successes and constraints. *Annual Review of Entomology* **40:** 559-585.

Cork A and Hall D R 1998. Application of pheromones for crop pest management in the Indian sub-continent. *Journal of Asia-Pacific Entomology* **1:** 35-49.

Critchley B R, Campion D G, Mc Veigh L J, Mc Veigh E M, Cavanagh G G, Hosny M M, Nasr E S A, Khidr A A and Naguib M 1985. Control of Pink boll worm *Pectinophora gossypiella* (Saunders) (Lepidoptera: Gelchiidae), in Egypt by mating disruption using hollow fibre laminate-flake and microencapsulated formulations of synthetic pheromone. *Bulletin of Entomological Research* **75:** 329-345.

Hall D R, Beevor P S, Campion D G, Chamberlain D J, Cork A, White R, Almetre A, Henneberry T J, Nandagopal V, Wightman J and Ranga Rao G V 1994. Identification and synthesis of new pheromones IOBC. *WPRS Bulletin* **16:** 1-10.

Hummel H E and Miller T A Eds 1984. Techniques in Pheromones research. New York: Springer-Verlag. 64pp.

Jutsum A R and Gordon R F S 1989. Pheromones: importance to insects and role in pest management. Insect pheromones in plant protection, AR Jutsum, and RFS Gordon.(eds.). John Wiley and Sons Ltd., New York, New York, 1-13.

Meerman F, Van De Ven G W J, Van Keulen H and Breman H 1996. Integrated Crop Management: An approach to sustainable agriculture and development. *International Pest Management* **42:** 13-14.

Metcalf R L 1998. Ultra microchemistry of Insect Semiochemicals. *Mickrochimica Acta* **129:** 167-180.

Mitchell E R 1981. Management of insect pests with semiochemicals. Concepts and practice Plemumn, New York/London pp, 514. Pheromone Research, Springer New York, pp. 111-129 Pheromone of *Sitona lineatus. Naturwissenschaften* **71:** 480-481.

Roelofs W L and Carde R T 1977. Responses of Lepidoptera to synthetic sex pheromone chemicals and other analogues. *Annual Review of Entomology* **22:** 377-399.

Rosca I, Brudea V, Bucurean E, Muresan F, Sandru I, Udrea A and Voicu M 1985. Possibilities of using synthetic sex pheromone in protection of cereal and technical crop cultures. In: Plant Protection, Proceedings of 9th National Conference Bucharest, 1985. *Acad. Stiinte Agric. Silv, Fundulea* **2:** 1-3.

Shanower T G, Wightman J A and Gutierrez A P 1993. Biology and control of groundnut leaf miner (*Aproaerema modicella* Deventor) (Lepidoptera: Gellichidae). *Crop Protection* **12:** 3-10.

Yadav J S, Chandrasekhar S and Rajasekhar K 1995. Short and stereoselective synthesis of pheromone components of *Aproaerema modicella. Synthetic Communications* **25:** 4035-4043.

Yadav J S, Prasad A R, Prasuna A L and Jyothi K N 2007. Potential of synthetic sex pheromone for the management of groundnut leaf miner *Aproaerema modicella* Deventer (Gelechiidae: Lepidoptera). *Journal of Applied Zoological Researches* **18:** 9-14.

Chapter 3

Biological Control of Soil-borne Plant Pathogens using Arbuscular Mycorrhizal Fungi

D J Bagyaraj

Center for Natural Biological Resources and
Community Development (CNBRCD),
41 RBI Colony, Anand Nagar, Bengaluru – 560 024, Karnataka
E-mail: djbagyaraj@gmail.com

ABSTRACT

The role of arbuscular mycorrhizal fungi (AMF) in improving plant growth is now well documented. Most of the studies on AMF-root pathogens suggest that AMF decreased or mitigated the disease severity. Consistent reduction of disease symptoms has been described for fungal, bacterial and nematode pathogens. Studies conducted so far suggest that the mechanisms of suppression may be due to morphological, physiological and biological alterations in the host. Thickening of the cell walls through lignification and production of other polysaccharides in mycorrhizal plants preventing penetration and growth of pathogens like Fusarium oxysporum *and* Phoma terrestris *have been demonstrated. Higher concentration of ortho-dihydroxy phenols present in mycorrhizal plants compared to non-mycorrhizal plants was found to be inhibitory to the root rot pathogen,* Sclerotium rolfsii. *The activation of specific plant defence mechanisms as a response of AMF colonization is an obvious basis for the protective capacity of AMF. Among the compounds involved in plant defense studied in relationship to AMF formation are phytoalexins, enzymes of the phenylpropanoid pathway, chitinases, peroxidases, pathogenesis related (PR) proteins etc. Mycorrhizal plants harbour higher population of microorganisms in the rhizosphere thus making it difficult for the pathogen to compete and gain access to the root. Further mycorhizosphere supports higher population of antagonists and siderophore producers. Thus the possibility of biologically controlling the root pathogens with AMF looks promising.*

Keywords: AM fungi, Biocontrol, Soil-borne plant pathogens.

Introduction

Plant roots provide an ecological niche for many of the microorganisms that abound in soil. German botanist Albert Bernard Frank in 1885 introduced the Greek word 'mycorrhiza' which literally means "fungus root". Most plant roots form mycorrhizal associations of one kind or the other with certain fungi in soil. These mycorrhizal fungi perform the function of root hairs. The symbiosis is so well balanced that, although many of the host cells are invaded by the fungal endophytes there is neither visible tissue damage nor low degree of pathogenicity towards its host. Mycorrhizal association generally enhances the growth and vigour of the host plants. Because of their wide spread occurrence in nature and their numerous benefits to plants, these fungi are currently attracting much attention in agricultural, horticultural and forestry research. Though there are different mycorrhizal associations, the most common type occurring in all ecological situations, especially in tropics, is the arbuscular mycorrhiza (AM) (Bagyaraj, 2006).

Arbuscular mycorrhiza are geographically ubiquitous and occur over a broad ecological range. They are commonly found in association with agricultural crops, most shrubs, most tropical tree species and some temperate tree species. AM has been observed in 1000 genera of plants representing some 200 families. There are atleast 300,000 respective hosts in the world flora and around 230 species of AM fungi (Bagyaraj, 2011). AM fungi belong to the phylum Glomeromycota, which has a single class Glomeromycetes with four orders Glomerales, Diversisporales, Paraglomerales and Archaeosporales. There are 11 families, 17 genera and 228 species. The commonly occurring genera of AM fungi are *Glomus*, *Gigaspora*, *Acaulospora*, *Entrophospora* and *Scutellospora*. The AM endophytes are not host specific, although evidence is growing that certain endophytes may form preferential association with certain host plants. These fungi are obligate biotrophs. Increased plant growth because of AM colonization is attributed to enhanced uptake of diffusion limited nutrients, hormone production, biological nitrogen fixation, drought resistance and suppression of root pathogens.

Biological control can be defined as the directed, accurate management of common components of ecosystems to protect plants against pathogens. Biological control of plant pathogens is currently accepted as a key practice in sustainable agriculture because it is based on the management of a natural resource, *i.e.*, certain rhizosphere organisms. Thus, biological control preserves environmental quality by reduction in chemical inputs and is characteristic of sustainable management practices (Barea and Jeffries, 1995). Several workers have reported that AM fungi can act as biocontrol agents for alleviating the severity of disease caused by root pathogenic fungi, bacteria and nematodes.

It is evident that an increased capacity for nutrient acquisition resulting from mycorrhizal association could help the resulting stronger plants to resist stress. However, AM symbiosis may also improve plant health through a more specific increase in protection (improved resistance and/or tolerance against biotic and abiotic stresses (Bethlenfalvay and Linderman, 1992; Barea and Jeffries, 1995). The study of a possible role for AM symbioses in protection against plant pathogens began in the 1970s, and a great deal of information has been published on the subject, however, the underlying mechanisms needs more investigation (Azcon-Aguilar and Barea, 1996). In this paper a brief account of AM fungi is given, followed by role played by them in controlling the root pathogens highlighting the mechanisms involved and suggesting the possible future line of work, wherever necessary.

Effect of AM Fungi on Root Pathogens

Most of the studies on AM fungi-root pathogens suggested that AM fungi decreased or mitigated the disease severity. Several review articles have appeared on this topic (Schenck, 1987; Hooker *et al.,* 1994; Linderman, 1994; Azcon-aguilar and Barea, 1996; Bagyaraj, 2006; St-Arnaud and Vujanovic, 2007; Tahat *et al.,* 2010). The main conclusions that can be drawn are: 1. AM associations can reduce damage caused by soil-borne plant pathogens, 2. The abilities of AM symbiosis to enhance resistance or tolerance in roots are not equal for the different AM fungi so far tested, 3. Protection is not effective for all pathogens, and 4. Protection is modulated by soil and other environmental conditions. Thus it can be expected that interactions between different AM fungi will vary with the host plant and the culture system. Recent studies have shown that AM fungi differ in their ability to protect plants against different soil-borne plant pathogens. The identity of AM fungal isolate has a dramatic effect on the level of protection against a particular pathogen (Wehner *et. al.,* 2010; Veresoglou and Rillig, 2012). Consistent reduction of disease symptoms has been described for fungal pathogens such as *Phytophthora parasitica, P. cactorum, P. vignae, Gaeumannomyces graminis* var. *tritici, Fusarium oxysporum, Chalara (Thielaviopsis) basicola, Rhizoctonia solani, R. bataticola, Sclerotium rolfsii, Pythium ultimum, P. splendens, Dothiorella gregania, Botrytis fabae, Ganoderma pseudoferreum* and *Aphanomyces* spp., bacteria such as *Pseudomonas syringae* and *Ralstonia solanacearum* and nematodes such as *Meloidogyne avenaria, M. incognita, M. hapla, M. javanica, Tylenchulus semipenetrans, T. vulgaris, Pratylenchus brachyurus, P. zeae, Helichotylenchus dihystera* and *Radopholus similis* (Rabie, 1998; Srivastava *et al.,* 2001; Bagyaraj, 2006; Tahat *et al.,* 2010). Further integrating AM fungi and castor cake in the management of root-knot nematode on egg plant has shown that seedlings colonized with mycorrhiza were least infected by *M. incognita* when transplanted in soil which was amended with castor cake. Significant increases in colonization of *Glomus fasciculatum* on roots of eggplant and chlamydospore densities of mycorrhiza indicated favourable effects of castor cake amendment on the growth of *G. fasciculatum* (Rao *et al.,* 1998).

Another study with staggered inoculation of AMF and root pathogen brought out that application of AM two weeks earlier than pathogen was better than simultaneous application of AM and the pathogen in reducing the severity of disease. Application of the pathogen two weeks earlier than AM was not effective in controlling the disease (Suresh and Bagyaraj, 1984). Vast (1997) studied the interaction of AM of *Arabica* coffee and the chemical control of nematodes. It was found that when AM fungi were inoculated well in advance, it resulted in an enhanced tolerance of *Arabica* coffee cultivars susceptible to endoparasitic nematodes of *Pratylenchus* and *Meloidogyne* that are wide spread in *Arabica* plantations.

Most of the AM-root pathogen interaction studies have been conducted in crop plants. But the information available on forest tree species is scanty (Mohan and Verma, 1996). There are a large number of root pathogens causing disease of forest tree species, especially in the nursery (Bagyaraj and Kehri, 2012). Availability of healthy stock seedling for planting in the field is an important aspect of forest management. Hence there is a need to carry out more studies on the role of AM fungi in the control of root pathogens of forest tree species.

Mechanisms of Suppression of Root Pathogens by AM Fungi

Studies conducted so far suggest that the mechanisms of suppression may be due to morphological, physiological and biological alterations in the host.

Morphological Alterations

It has been demonstrated that AM colonization induces remarkable changes in root system morphology, as well as in the meristematic and nuclear activities of root cells (Atkinson *et al.*, 1994). This might affect rhizosphere interactions and particularly pathogen-infection development. The most frequent consequence of AM colonization is an increase in branching, resulting in a relatively larger proportion of higher order roots in the root system (Hooker *et al.*, 1994). However, the significance of this finding for plant protection has not yet been sufficiently considered. A recent study brought out that root architecture plays an important role in AM fungi protecting plants against root pathogens (Sikes, 2010). More attention needs to be given to root system morphology in the future because it could modify the infection dynamics of the pathogen as well as the pattern of resistance of AM roots to pathogen attack.

Thickening of the cell walls through lignification and production of other polysaccharides in mycorrhizal plants preventing penetration and growth of pathogens like *Fusarium oxysporum* and *Phoma terrestris* have been demonstrated. A stronger vascular system observed in mycorrhizal plants will increase the flow of nutrients, impart greater mechanical strength and diminish the effect of vascular pathogens. Histopathological studies of nematode galls caused by *Meloidogyne* spp. showed that galls in mycorrhizal plants had fewer giant cells or syncitia, which are needed for the development of nematode larvae, compared to non-mycorrhizal plants. The nematodes in mycorrhizal plants were smaller and took a longer time to develop into adults (Bagyaraj, 1996).

Pioneering observations by Dehne (1982) illustrated how fungal root pathogens and AM fungi, although colonizing the same host tissues, usually develop in different root cortical cells, indicating some sort of competition for space. Cordier *et al.* (1996) working with AM-*Phytophthora* interaction observed that the pathogen does not penetrate cortical cells containing arbuscule; suggesting that localized competition for infection/colonization site does occur between the pathogen and the AM fungus. Contrary to this in a study, epicotyls of mycorrhizal pea plants showed a reduction in disease severity caused by *Aphanomyces eutriches* although this part of the plants was not mycorrhizal. This suggested that an induced systemic factor may be responsible for increased resistance in mycorrhizal plants (Bodker *et al.*, 1998).

Physiological and Biochemical Alterations

Improved Nutrient Status of the Host Plant

Increased nutrient uptake made possible by AM symbiosis results in more vigorous plants. The plant itself may thus be more resistant to or tolerant of pathogen attack though this is true, later studies using non-mycorrhizal plants with tissue P concentration similar to that of mycorrhizal plants, indicated that non-P mediated mechanisms are also involved (Bodker *et al.*, 1998).

Competition for Host Photosynthates

It has been proposed that the growth of both the AM fungi and root pathogens depends on host photosynthates and that they compete for the carbon compounds reaching the root (Linderman, 1994). When AM fungi have primary access to photosynthates, the higher carbon demand may inhibit pathogen growth. However, there is little or no evidence that competition for carbon compounds is a generalized mechanism for pathogen biocontrol activity of AM symbiosis. Colonization of a plant by AM fungi alters the host physiology and in turn the root exudation pattern (Mada and Bagyaraj, 1993). Decreased root exudation in mycorrhizal plants increased membrane phospholipid content, possibly helps in reducing the infection by root pathogens.

Higher Levels of Phenols and Amino Acids

Higher concentration of ortho-dihydroxy phenols present in mycorrhizal plants compared to non-mycorrhizal plants was found to be inhibitory to the root-rot pathogen *Sclerotium rolfsii*. Higher levels of amino acid arginine in the root extracts of mycorrhizal plants were found to reduce *Thielaviopsis basicola* chlamydospore production. Increased phenylalanine and serine in tomato roots due to inoculation with *G. fasciculatum* was found to be inhibitory to root knot nematodes (Bagyaraj, 2006).

Activation of Plant Defence Mechanisms

During their life cycle, plants evolve a number of defence responses elicited by various signals, including those associated with pathogen attack (Huynh *et al.*, 1992). The activation of specific plant defence mechanisms as a response to AM colonization is an obvious basis for the protective capacity of AM fungi. Among the compounds involved in plant defence (Bowles, 1990) studied in relationship to AM formation are phytoalexins, enzymes of the phenylpropanoid pathway, chitinases, β-1,3-glucanases, peroxidases, pathogenesis related proteins, callose, hydroxyproline rich glycoproteins (HRGP) and phenolics (Gianinazzi Pearson *et al.*, 1994). A recent study showed that application of a consortium of five AM fungi enhanced polyphenol oxidase, phenylalanine ammonium-lyase and peroxidases in the common bean plant and thus protected it against *Fusarium* wilt (Al-Askar and Rashad, 2010).

Phytoalexins, low-molecular-weight, toxic compounds usually accumulating with pathogen attack and released at the sites of infection, were detected during the later stages of AM formation (Morandi *et al.*, 1984). There seems to be a similar low activation of the phenylpropanoid-related enzymes. In particular, both phenylalanine ammonium-lyase, the first enzyme of the phenylpropanoid pathway, and the chalcon isomerase, the second enzyme specific for flavanoid/isoflavanoid biosynthesis, increased in amount and activity during early colonization of plant roots by *Glomus intraradices*, but then decreased sharply to levels at or below those in uninoculated controls. (Volpin *et al.*, 1995). These results suggest that AM fungi initiate a host defence response, which is subsequently suppressed. Pathogenesis related proteins and HRGP are synthesized only locally and in very low amount in response to AM colonization.

Electrophoretic analysis of soluble extracts from AM roots has demonstrated that the host plant produces a number of new proteins (endomycorrhizins) in response to AM colonization (Gianinazzi pearson and Gianinazzi, 1995). New polypeptides are synthesized during AM infection and others disappear (Dumas Gaudot *et al.*, 1994). However, this altered pattern of protein synthesis in the plant is not necessarily related to defence reactions. This is a research area deserving further attention. It can be said that plant defence mechanisms are activated to a very little extent by AM colonization. However, these compounds could sensitize the roots to pathogens and enhance mechanisms of defence to subsequent pathogen infection (Azcon Aguilar and Barea, 1996; Bagyaraj, 2006). Current research using molecular biology techniques and immunological and histochemical analyses will probably provide more information about these mechanisms.

Biological Alterations

Mycorrhizal plants harbour higher population of microorganisms in the rhizosphere thus making it difficult for the pathogen to compete and gain access to the root. Caron (1989) reported a reduction in *Fusarium* population in the soil surrounding mycorrhizal tomato roots as compared with the soil of non-mycorrhizal controls. Secilia and Bagyaraj (1987) isolated more pathogen-antagonistic actinomycetes from the rhizosphere of AM plants than from non-mycorrhizal controls, an effect that also depended on the AM fungus involved. Microorganisms producing siderophores, which are low molecular weight chelating agents that have high affinity for ferric iron and thus fungistatic to many pathogens, were observed in higher numbers in the rhizosphere of mycorrhizal plants. The prophylactic ability of some AM fungi could be exploited in association with other rhizosphere microorganisms known to be antagonistic to root pathogens that are being used as biological control agents (Azcon Aguilar and Barea, 1996).

Earlier studies suggest that microbial antagonists of fungal pathogens, either fungi or plant growth promoting rhizomicroorganisms (PGPR) do not antagonize AM fungi. Moreover, they can improve the development of the mycosymbiont and facilitate AM formation (Lioussanne *et al.*, 2009). Synergistic interactions between AM fungi and PGPR or other soil microorganisms may enhance bioprotection against soil-borne plant diseases. For example, Siddiqui and Mahmood (1998) demonstrated that the combination inoculation of *G. mosseae* and *P. fluorescens* caused a greater reduction in galling and nematode reproduction than when they were used alone. The mechanisms used by PGPR to protect plants against pathogens are well known: competition for space and nutrients, modification of Fe and Mn availability, liberation of antibiotics and HCN, plant growth promotion by modification of plant hormone balance, and stimulation of systemic and localized plant defence mechanisms (Lioussanne *et al.*, 2009).

Certain microorganisms have recently been shown to stimulate colonization by AM fungi. These have been called mycorrhiza helper organisms (MHO) (Jayanthi *et al.*, 2003). Some of these MHOs are also PGPRs. Co-inoculation of MHOs along with AM fungi protecting the plants better than AM fungus alone against root pathogens has been reported recently by some workers. *Paenibacillus* sp. strain B2 along with *G. mosseae* protected sorghum plants more effectively against the pathogen *Phytophthora nicotianae* compared to *G. mosseae* alone (Budi *et al.*, 2000). Wilt disease complex of the medicinal plant *Coleus forskohlii* caused by *Fusarium chlamydosporum*, *Pseudomonas fluorescens* and *Meloidogyne incognita* is very serious in India. Inoculation with the AM fungus *G. fasciculatum* together with *P. fluorescens* and *Trichoderma viride* was found to increase root yield and root forskolin concentration, and reduce the severity of the disease significantly under field conditions compared to inoculation with only *G. fasciculatum* (Singh *et al.*, 2012). This brings out that microbial consortia consisting of effective AM fungi and PGPRs/MHOs can be judiciously selected for controlling soil-borne plant diseases. Therefore, the management of these interactions improving plant growth and health, in an integrated approach, should be one of the targets for future research.

It can be concluded that AM fungi has the potential to alleviate the severity of disease caused by soil-borne plant pathogens. Like most instances of biological control AM fungi cannot offer complete immunity against the soil-borne plant pathogens. They could only impart a degree of resistance against these pathogens. The diversity of interactions between AM fungi and soil-borne plant pathogens show that each pathogen-AM fungi-plant combination is unique which is further influenced by the environment. However, the possibility of biologically controlling these pathogens with AM fungi alone or together with PGPRs/MHOs looks promising. Most of the soil-borne pathogens are currently controlled only with expensive physical or chemical soil treatments. Thus, AM fungi offer an alternative approach and we should pursue their potential as biocontrol agents. Further, most of the studies, barring few investigations, have been conducted in pots and hence there is a need for exploiting this potential under field conditions.

References

Al-Askar A A and Rashad Y M 2010. Arbuscular mycorrhizal fungi: a biocontrol agent against common bean *Fusarium* root rot disease. *Plant Pathology Journal* 9: 31-38.

Atkinson D, Berta G and Hooker J E 1994. Impact of mycorrhizal colonization on root longevity and the formation of growth regulators. In: *Impact of Arbuscular Mycorrhizas on Sustainable Agriculture and Natural Ecosystems* (eds. Gianinazzi S and Schuepp H), Birkhauser, Basel, pp 89-99.

Azcon Aguilar C and Barea J M 1996. Arbuscular mycorrhizas and biological control of soilborne plant pathogens-an overview of the mechanisms involved. *Mycorrhiza* **6:** 220-223.

Bagyaraj D J 1996. Mycorrhizal symbiosis in tropical trees in relation to control of root pathogens. In: *Impact of Diseases and Insect Pests in Tropical Forests,* Kerala Forest Research Institute, Peechi, India. pp. 239-245.

Bagyaraj D J 2006. Current status of VAM as biocontrol agents for disease management. In: Current Status of Biological Control of Plant Diseases Using Antagonistic Organisms in India (eds. Ramanujam B and Rabindra R J), Project Directorate of Biological Control Pub., Bangalore. pp. 125-134.

Bagyaraj D J 2011. Microbial Biotechnology for Sustainable Agriculture, Horticulture and Forestry, New India Publishing Agency, New Delhi. 308 pp.

Bagyaraj D J and Kehri H K 2012. AM fungi: Importance, nursery inoculation and performance after outplanting. In: *Microbial Diversity and Functions* (eds. Bagyaraj D J, Tilak K V B R and Kehri H K), New India Publishing Agency, New Delhi. pp. 641-668.

Barea J M and Jeffries P 1995. Arbuscular mycorrhizas in sustainable soil plant systems. In: Mycorrhiza-Structure, Function, Molecular Biology and Biotechnology (eds. Varma A and Hock B), Springer, Heidelberg. pp. 521-529.

Bethlenfalvay G J and Linderman R G 1992. Mycorrhizae in Sustainable Agriculture, American Society of Agronomy, Inc., Madison, Wisconsin. 124 pp.

Bodker L, Kjoller R, Rosendahl S 1998. Effect of phosphate and the arbuscular mycorrhizal fungus *Glomus intraradices* on disease severity of root-rot of peas (*Pisum sativum*) caused by *Aphanomyces euteiches*. *Mycorrhiza* **8:** 169-174.

Bowles D J 1990. Defense related proteins in higher plants. *Annual Review of Biochemistry* **59:** 873-907.

Budi S W, van Tuinen D, Arnould C, Dumas-Gaudut E, Gianinazzi-Pearson V and Gianinazzi S 2000. Hydrolytic enzyme activity of *Paenibacillus* sp. strain B2 and effect of antagonistic bacterium on cell wall integrity of two soil-borne pathogenic fungi. *Applied Soil Ecology* **15:** 191-199.

Caron M 1989. Potential use of mycorrhizae in control of soil-borne diseases. *Canadian Journal of Plant Pathology* **11:** 177-179.

Cordier C, Gianinazzi S and Gianinnazzi-Pearson V 1996. Colonization patterns of root tissues by *Phytophthora nicotiana* var. parasitica related to reduced disease in mycorrhizal tomato. *Plant and Soil* **185:** 223-232.

Dehne H W 1982. Interaction between vesicular-arbuscular mycorrhizal fungi and plant pathogens. *Phytopathology* **72:** 1115-1119.

Dumas Gaudot E, Guillaume P, Tahiri-Alaoui A, Gianinazzi-Pearson V and Gianinazzi S 1994. Changes in polypeptide patterns in tobacco roots colonized by two *Glomus* species. *Mycorrhiza* **4:** 215-221.

Gianinazzi Pearson V, Gollotte A, Dumas Gaudot E, Franken P and Gianiazzi S 1994. Gene expression and molecular modifications associated with plant responses to infection by arbuscular mycorrhizal fungi. In: *Advances in Molecular Genetics of Plant Microbe Interactions* (eds. Hennecke H and Verma D P S), Kluwer, Dordsecht. pp. 179-186.

Gianinazzi-Pearson V and Gianinazzi S 1995. Proteins and protein activities in endomycorrhizal symbiosis. In: *Mycorrhiza-Structure, Function, Molecular Biology and Biotechnology* (eds. Varma A and Hock B), Springer, Heidelberg. pp. 251-266.

Hooker J E, Jaizme-Vega M and Atkinson D 1994. Biocontrol of plant pathogens using arbuscular mycorrhizal fungi. In: *Impact of Arbuscular Mycorrhizas on Sustainable Agriculture and Natural Ecosystems* (eds. Gianinazzi S. and Schuepp H), Birkhauser, Basel, pp. 191-200.

Huynh Q K, Hironara C M, Levine E B, Smith C E, Borgmeyer J R and Shah D M 1992. Antifungal proteins from plants. Purification, molecular cloning and cloning and antifungal properties of chitinases in maize seeds. *Journal of Biology and Chemistry* **26:** 6635-6640.

Jayanthi Srinath, Bagyaraj D J, and Sathyanaraya B N 2003. Enhanced growth and nutrition of micropropagated *Ficus benjamina* to *Glomus mosseae* co-inoculated with *Trichoderma harzianum* and *Bacillus coagulans*. *World Journal of Microbiology and Biotechnology* **19:** 69-72.

Linderman R G 1994. Role of VAM fungi in biocontrol. In: *Mycorrhizae and Plant Health* (eds. Pfleger F L and Linderman R G), ASP, St. Paul. pp. 1-26.

Lioussanne L, Beauregard M S, Hamel C, Jolicoeur M and St-Arnaud M 2009. Interactions between arbuscular mycorrhizal fungi and soil microorganisms. In: *Advances in Mycorrhizal Science and Technology* (eds. Khasa D, Piche Y and Coughlan A P), NRC, Research Press, Ottawa. pp. 51-70.

Mada R J and Bagyaraj D J 1993. Root exudation from *Leucaena leucocephala* in relation to mycorrhizal colonization. *World Journal of Microbiology and Biotechnology* **9:** 342-344.

Mohan V and Verma N 1996. Interactions of VA-mycorrhizal fungi with rhizosphere and rhizoplane mycoflora of forest tree species in arid and semi arid regions. In: *Impact of Diseases and Insect Pests in Tropical Forests*, Kerala Forest Research Institute, Peechi, India. pp. 222-231.

Morandi D, Baily J A and Gianinazzi-Pearson V 1984. Isoflavonoid accumulation in soybean roots infected with vesicular-arbuscular mycorrhizal fungi. *Physiological Plant Pathology* **24:** 357-364.

Rabie G H 1998. Induction of fungal disease resistance in *Vicia faba* by dual inoculation with *Rhizobium leguminosarum* and vesicular-arbuscular mycorrhizal fungi. *Mycopathologia* **141:** 159-166.

Roa M S, Reddy P P, Sukhada M, Nagesh M and Pankaj 1998. Management of root-knot nematode on egg plant by integrating endomycorrhiza (*Glomus* *fasciculatum*) and castor (*Ricinus communis*) cake. *Nematologia-Mediterranea* **26**: 217-219.

Schenck N C 1987. Vesicular-arbuscular mycorrhizal fungi and the control of fungal root disease. In: *Innovative Approaches to Plant Disease Control* (ed Chet I), Wiley, New York. pp. 179-191.

Secilia J and Bagyaraj D J 1987. Bacteria and actinomycetes associated with pot cultures of vesicular arbuscular mycorrhizas. *Canadian Journal of Microbiology* **33**: 1069-1073.

Siddiqui Z A and Mahmood I 1998. Effect of a plant growth promoting bacterium, an AM fungus and soil types on the morphometrics and reproduction of *Meloidogyne javanica* on tomato. *Applied Soil Ecology* **8**: 77-84.

Sikes B A 2010. When do arbuscular mycorrhizal fungi protect plant roots from pathogens? *Plant Signal Behaviour* **5**: 763-765.

Singh R, Kalra A, Ravish B S, Divya S, Parameswaran T N, Srinivas K V N S and Bagyaraj D J 2012. Effect of potential bioinoculants and organic manures on root-rot and wilt, growth, yield and quality of organically grown *Coleus forskohlii* in a semi arid tropical region of Bangalore (India). *Plant Pathology* **61**: 700-708.

Srivastava A K, Ahmed R, Kumar S, Sukhada Mohandas 2001. Role of VA mycorrhiza in the management of Guava in the alfisols of Chotanagpur. *Indian Phytopathology* **54**: 32-34.

St-Arnaud M and Vujanovic V 2007. Effects of the arbuscular mycorrhizal symbiosis on plant diseases and pests. In: *Mycorrhizae in Crop Production* (eds. Hammel C and Plenchette C), Haworth Press Inc., New York. pp. 67-122.

Suresh C K and Bagyaraj D J 1984. Interaction between a vesicular arbuscular mycorrhiza and root knot nematodes and its effect on growth and chemical composition of tomatoes. *Nematologia Mediterranea* **12**: 31-39.

Tahat M M, Kamaruzaman, Sijan and Othman R 2010. Mycorrhizal fungi as biocontrol agent. *Plant Pathology Journal* **9**: 198-207.

Vast P 1997. Interaction of endomycorrhizae of Arabica coffee and the chemical control of nematodes (*Pratylenchus coffeae* and *Meloidogyne konaensis*) *Dix-septieme colloque Scientifique International surle Café*. pp. 564-571.

Veresoglou S D and Rillig M C 2012. Suppression of fungal and nematode plant pathogens through arbuscular mycorrhizal fungi. *Biology Letters* **8**: 214-217.

Volpin H, Phillips D A, Okon Y and Kapulnik Y 1995. Suppression of an isoflavanoid phytoalexins defence response in mycorrhizal alfalfa roots. *Plant Physiology* **108**: 1449-1454.

Wehner J, Antunes P M, Powell J R, Mazukatow and Rillig M C 2010. Plant pathogen protection by arbuscular mycorrhizas: A role for fungal diversity *Pedobiologia* **53**: 197-201.

Chapter 4

Addressing Plant Health Management using Diagnostics and Certification Issues

V Celia Chalam[1], R K Khetarpal[2] and P C Agarwal[1]

*[1]Division of Plant Quarantine, ICAR-National Bureau of Plant Genetic Resources,
Pusa Campus, New Delhi – 110 012
[2]CABI South Asia, DP Shastri Marg, NASC Complex,
Pusa, New Delhi – 110 012
E-mail: mailcelia@gmail.com, celia@nbpgr.ernet.in*

ABSTRACT

Emerging, re-emerging and endemic pests continue to challenge our ability to safeguard plant health and can be managed most effectively if pest-free certified planting material is used and control measures are introduced at an early stage. Early, sensitive and accurate diagnosis is necessary for application of mitigation strategies. Biological techniques for pest detection are usually highly accurate but too slow and not amenable to large-scale application. DNA-based technologies improve the speed and accuracy of detection of all groups of pests. Comparison of ITS regions of bacteria, fungi, and nematodes has proven useful for taxonomic purposes. Sequencing of conserved genes has been used to develop PCR-based detection with varying levels of specificity for viruses, fungi, bacteria, etc. Combinations of ELISA and PCR technologies are used to improve sensitivity of detection and to avoid problems during PCR with inhibitors. Attention is now given to techniques like Multiplex-PCR, Real-time PCR, LAMP, HDA, etc. LAMP and HDA are isothermal DNA methods, which do not require a thermal cycler and has potential applications in breeding, certification, quarantine, biosecurity and microbial forensics. Widespread adoption of SOPs and diagnostic laboratory accreditation serve to build confidence among institutions. Certification is an important means of managing pests and is imperative since inadvertent use of infected seeds and other planting material will not only result in its poor performance, but also in deleterious spread of pests. In India, the Seeds Act, 1966 (including the Seed Bill 2004), does not require a mandatory seed certification against any pests. Out of 110 crops for which seed certification standards are prescribed, seed health standards by seed analysis are available only for nine crops. About 130 plant viruses are known to be seed-transmitted, but there are no seed health standards prescribed for viral diseases at seed stage. Seed analysis is carried out essentially by dry seed examination, though many advanced detection techniques are available. Recently, the seed standards for certification against viruses have been proposed, based on virus spread using a known level of initial seed/seedling infection. National Certification System for Tissue Culture-raised Plants (NCS-TCP) has been developed for the first time in the world by DBT, Government of India, where currently no such organized structure exists for certification of tissue culture material. Also, the planting material under international exchange needs to be certified as pest-free to minimize the risk associated with the introduction of exotic pests. Similarly, conservation of pest-free seeds and in-vitro cultures of a crop in Gene Banks will minimize the spread of "germplasm-borne" pests. The quality control and certification approach for pests makes it imperative that India should put in place a National Plant Pests Diagnostic and certification Network. The certification would be useful only if it is simultaneously ensured that commercial planting material, germplasm and breeding material are also free from all groups of pests. The efforts for the certification approach would also be worthwhile only if the importance of pests in trade and exchange is properly appreciated by both the public and private sectors.

Keywords: Certification, Seed health Standards; Virus diagnostics.

Introduction

Plant diseases and insect pests substantially reduce crop production every year, resulting in massive economic losses throughout the world. Historically, crop epidemics of diseases coupled with other factors have even led to certain famine-like situations. The famous Irish potato famine caused by potato late blight disease in 1845 resulted in death of approximately 1 million people and emigration of a million more from Ireland. The rice blast was responsible for famine in Japan during the 1930s, and brown spot of rice for the famous Bengal famine in India in 1943.

Management of diseases and insect pests is of utmost importance for sustaining food production. Plant health management is the science and practice of understanding and overcoming the succession of biotic and abiotic factors that limit plants from achieving their full genetic potential (Cook, 2000). The most fundamental approach to the management of a disease or insect pest is to ensure that it is not present through exclusion (quarantine) or eradication. National Plant Protection Services assume responsibility for protecting their countries from the unwanted entry of new pathogens and insect pests and for coordinating programmes to eradicate those that have recently arrived and are still sufficiently confined for their elimination to be realistic. The strategies for plant health management include: certified disease-free seed and other planting material, chemical control, biological control, cultural control, and use of resistant varieties.

Emerging, re-emerging and endemic pests continue to challenge our ability to safeguard plant health and can be managed most effectively if pest-free certified seed and other planting material is used and control measures are introduced at an early stage. Early, sensitive and accurate diagnosis is indispensable for application of mitigation strategies. Since both seeds and vegetative propagules (such as bulbs, rhizomes, suckers, rooted cuttings, etc.) are efficient carriers of pests especially viruses, infection is progressively transmitted through generations. Further, if the *in vitro* cultures are raised from a virus-infected mother plant, the plantlets are bound to carry the virus infection. The selection of a diagnostic method for evaluating plant health depends on the host to be tested and the type of pests that may be carried in the seed/vegetative propagules/*in vitro* cultures. The technique should be reliable for field performance and quarantine requirements, reproducible within statistical limits, economical with regard to time, labour and equipment and should be rapid to provide results of large samples in the shortest time.

In the context of quality control, bulk samples of seed lots/vegetative propagules/tissue culture-raised plants need to be tested by drawing workable samples as per norms. The detection of pests is then carried out by the approved or available techniques. Over the years a great variety of methods have been developed that permits the detection and identification of pests.

Diagnostics for Detection of Pests

The successful detection and control of pests in seed and other planting material depend upon the availability of rapid, reliable, robust, specific and sensitive methods for detection and identification of pests. The various techniques, conventional and modern, that are employed for detection of pests are enumerated below in brief:

I. Conventional Methods

The conventional techniques which are adapted are briefly described below:

i. Examination of dry seed

Examination of dry seed under a low power stereomicroscope with magnification upto 50 to 60 times may reveal symptoms such as discolouration, malformation, fruiting bodies of fungi, hyphae, eggs of insects on seed surface, insect feeding holes and even bacterial spores or growth on the surface of the seed.

ii. X-ray radiography of dry seed

Seeds of about 344 plant genera reported to harbour hidden infestation when exposed to soft X-rays at 22kv, 3mA for 15 seconds at the distance of 30 cm clearly indicate the presence of infestation caused by bruchids and phytophagous chalcids (Bhalla *et al.*, 2006). The technique is however, mainly used by quarantine stations.

iii. Transparency test

Small seeds can be rendered transparent by heating in lactophenol-acid fuchsin (phenol 2 parts, lactic acid 2 parts, glycerine 1 part, distilled water (hot) 2 parts, few drops of fuchsin) for detection of internal infestation by insects.

iv. Examination of seed washing

This technique is employed for detecting various fungi, adhered to the surface of seed and also the spore load in a short time. Weighed seeds are shaken in a known volume of water for a fixed time on a mechanical shaker. The washing is examined under a compound microscope. By the use of phase contrast microscope, unstained bacteria can also be observed. Failure to detect infection internal to the seed and inability to distinguish between spores of saprophytic fungi from spores of pathogenic fungi are the limitations of this technique.

v. Examination of soaked seed

This technique is practiced for detecting paddy bunt fungus, *Neovossia horrida*. Rice seeds are soaked in 0.2 per cent sodium hydroxide for 24 h at 25°C. The infected seeds appear brown, dull or shiny black. The infection is confirmed by rupturing the seed in a drop of water. A stream of smut spores are released from the shiny black discolouration.

vi. Examination of whole embryo

This technique is used for detecting obligate pathogens *i.e. Ustilago segetum* var. *hordei* and *U. segetum* var. *tritici* (loose smut of barley and wheat). Seeds are soaked overnight in 10 per cent sodium hydroxide, containing trypan blue stain at 25°C, washed with warm water through sieves of decreasing size and embryos are finally cleared in lactophenol. The infected embryos under stereomicroscope reveal bluish stained mycelium which may be present in scutellum, plumule bud or the whole embryo.

vii. Incubation tests

Incubation tests are used successfully against surface-borne as well as internal infections. After plating of seed, the incubation period gives an opportunity to the dormant mycelium of fungal spores to grow along with the host. Most commonly used incubation tests are blotter method and agar plate method, in which seeds are placed on moist blotters and agar media, respectively.

viii. Water-agar seedling symptom test

Seeds are sown singly on water agar, in test tubes or sometimes in microculture plastic plates or Petri dishes. They are incubated under 12 h alternating cycle of light and darkness. Seedlings are inspected for symptoms. This is an economical procedure for separating healthy and infected seedlings and has been extensively used for detection of many fungi in cucurbits and sesame.

ix. Phage sensitivity test

Most of the bacteria are sensitive to bacteriophages and as a result they are lysed. Formation of lytic zones around the phage spot confirms the identity/presence of the bacterium.

x. Staining of inclusion bodies

Inclusion bodies are aggregate of virus particles or virus induced proteins or special structures characteristic of virus infection either in the cytoplasm or the nucleus. The virus infection is detected by cutting free-hand sections with a razor blade and staining with Azure A and O-G combination. The tissue is observed under light microscope and inclusion bodies are located and characterized.

xi. Electron microscopy

The Transmission Electron Microscope (TEM) can be used directly to detect the presence of virus in the plant tissue. It reveals the shape and size of the virus particle. The shape and size of the virus particle gives an idea of the group to which it may belong. This helps in limiting the number of antisera to be used in serological tests such as Enzyme linked immunosorbent assay (ELISA) for further identification of the virus, as only antisera of viruses of a particular shape can thus be used for identification (Chalam and Khetarpal, 2008). However, the TEM remains very expensive equipment and is often not available.

xii. Growing-on test

Certain seed-borne diseases need longer periods for their expression than provided in the normal incubation tests. The pathogens are identified based on symptoms followed by tests of infectivity/electron microscopy/ELISA in case of viruses.

xiii. Infectivity test

Healthy young seedlings or mature plants are exposed to infected material to produce the symptoms. This approach is quite old; however, the technique has been used successfully to detect fungi, bacteria and viruses. In case of viruses, their presence is assayed by inoculating leaf extracts of seedlings, which may or may not be showing symptoms, on indicator hosts. The indicator hosts may reveal the symptoms by producing local lesions or systemic infection. Long span of time required for development of symptoms on them, requirement of large green house space and the prodigious labour and time for working with large samples are the limitations of this test.

II. Serological Tests/Immunoassays

This technique is based on the principle that a substance having high molecular weight (>10,000 daltons) when introduced into an animal, causes the formation of specific proteins (the immunoglobulins) in the blood, which are commonly called antibodies. The causative substance is called antigen and the blood serum containing antibodies is called antiserum. The antigen-antibody reaction can be examined *in vitro* as well as *in situ*. Serodiagnostic tests are very sensitive and reliable to detect the presence of virus and bacteria.

Besides, **Immunosorbent electron microscopy (ISEM)** is another sensitive technique coupling serology with electron microscopy used for detection of viruses. In this case, the electron microscope grids are pre-treated with specific antiserum, which facilitates the adherence of virus particles on the grid by several folds. The virus particles are then specifically trapped and decorated with the antiserum and the contaminants are washed away. ISEM is not well suited for testing of large number of samples but for small number of samples it can be much quicker than ELISA. The non-availability of the electron microscope and the antisera are often the limiting factors in adopting ISEM.

Earlier, serological tests based on immunoprecipitation, immunodiffusion, and latex agglutination were very popular. Now-a-days, ELISA and Dot immunobinding assay (DIBA) are the most widely used methods of serological detection of plant viruses and also bacteria as they are much more sensitive than diffusion and agglutination methods, use less antibody and can be employed for simultaneous handling of a large number of samples in routine testing. Thirty-four seed-transmitted viruses which are either not known to occur in India or are known to possess virulent strains or not known to occur in India on particular host(s), have been intercepted in germplasm including transgenics imported from many countries (Chalam and Khetarpal, 2008; Chalam *et al.*, 2009).

i. **Enzyme-linked immunosorbent assay (ELISA):** The adoption of ELISA test has created new interest in serological diagnosis of plant viruses. Two types of ELISA commonly used are: (i) Double antibody sandwich- ELISA (DAS-ELISA) and (ii) Direct antigen coating- ELISA (DAC-ELISA).

    The advantages of ELISA are it is reasonably sensitive; less susceptible to 'false positives'; low cost per sample; can handle large number of samples; can be subjected to automation and detection kits are available commercially.

ii. **Dot-immunobinding assay (DIBA):** It is a variant of ELISA test used for detection of viruses wherein instead of using microtitre plates as solid support, nitrocellulose membranes are used. A few microlitres of extract of infected samples are blotted on this membrane which is then submerged in primary antibody (crude antiserum). The precipitated antibody is then detected with enzyme-labeled second antibody.

    DIBA has an edge over conventional ELISA as it does not require any special equipment, it requires only a crude specific antiserum to each of the viruses/bacteria and a single enzyme conjugate. Above all, the blotted membranes can be mailed to long distances for further processing in a centralized laboratory.

iii. **Tissue blotting immunoassay/Tissue print immunoassay/Tissue print immunoblotting:** Tissue blot immunoassay is a detection method similar to DIBA, except that it does not involve tissue extraction. Instead of dotted extracts, freshly cut tissue surfaces are printed directly onto nitrocellulose membranes. The antigens trapped in the tissue blots are then reacted with antibodies, conjugate and substrate in the same way as in DIBA. The method has gained popularity for many purposes, not only due to its simplicity, eliminating the need for an extraction step, but also due to its high sensitivity, comparable with or in some cases even higher than ELISA and DIBA.

iv. **Lateral flow strip method:** Lateral flow strip method is a variation of ELISA used for detecting viruses/bacteria, and the antibodies are immobilised onto a test strip in specific zones. The test is provided in kit form and does not require any major equipment. Lateral flow strips are suitable for field or on-site use, with minimal training required. Sample preparation simply involves crushing the sample and mixing it with the extraction solution provided in the kit. These tests generally provide qualitative or semi-quantitative results using antibodies and colour reagents incorporated into a flow strip.

III. Molecular Methods

i. Polymerase chain reaction

This involves rapid and highly specific *in vitro* amplification of selected DNA sequences, for which specific primers are synthesized. With its relative simplicity and high sensitivity (detecting picogram quantities of viral nucleic acids in infected tissues), PCR method has high potential in detection of viruses/bacteria/nematodes/insect pests. However, the pre-requisite of having known sequences to select and synthesize suitable primers limits its application to well-characterized viruses/bacteria/nematodes/insect pests.

Variants of PCR

(a) **Reverse transcription PCR (RT-PCR):** Most of the plant viruses consists of RNA, which require the introduction of a preliminary reverse transcription (RT) step before the PCR amplification process (RT-PCR), thus allowing the amplification of RNA sequences in a cDNA form. Many viruses have been detected using RT-PCR (Chalam *et al.*, 2004, 2012a,b; Chalam and Khetarpal, 2008).

(b) **Immunocapture-PCR (IC-PCR):** This is a variant of PCR, which utilizes antibodies to trap viral particles without prior viral RNA extraction, which would presumably facilitate its use in routine testing. Moreover, because antibodies are involved in the first step, it may be assumed that, this method could also selectively detect viruses. Thus, this method could be very useful and practical in virus indexing programme.

The advantages of PCR are it is highly sensitive (can detect picogram quantities of target nucleic acid); the process is automated: very rapid, it takes 2 h or less for the test; can be used for detecting RNA or DNA and very useful where ELISA is not effective (viroids, geminiviruses).

(c) **Real time PCR/Real time RT-PCR:** Real-time PCR monitors the fluorescence emitted during the reaction as an indicator of amplicon production during each PCR cycle (*i.e.*, in real time) as opposed to the endpoint detection. The real-time progress of the reaction can be viewed. Real-time PCR quantitation eliminates post-PCR processing of PCR products. This helps to increase throughput and reduces the chances of carryover contamination. No-post PCR processing (no electrophoretical separation of amplified DNA) is required.

The advantages of Real-time PCR/Real-time RT-PCR are: it is not influenced by non-specific amplification; amplification can be monitored real-time; no post-PCR processing of products (high throughput, low contamination risk); ultra-rapid cycling (30 minutes to 2 hours); requires 1000-fold less RNA than conventional assays and is most specific, sensitive and reproducible. The technique has been successfully exploited for detecting viruses (Chalam *et al.*, 2004, 2012a) and bacteria.

ii. Nucleic acid hybridization assays

Detection of viruses/bacteria by nucleic acid hybridization is based on the specific pairing between the target nucleic acid sequence (denatured DNA or RNA) and a complementary nucleic acid probe to form double stranded nucleic acids. Thus, either RNA or DNA sequences may be used as probes. It has a potential of detecting extremely low level of inoculum or latent infections in plant/planting materials. Basically the method involves the immobilization of a spot or dot of sap extract from the plant under test on a solid matrix and the detection of viral/bacterial nucleic acid sequences in that spot by use of a hybridization probe.

iii. Double stranded RNA (dsRNA) analysis

For poorly characterized or unknown viruses detection methods either are not available or are not sensitive. In such cases, double-stranded RNA analysis is a rapid tool that can supplement the information obtained from bio-assays. Analysis of dsRNA is based on the isolation of disease specific dsRNAs from virus-infected tissues and their electrophoretic separation on a gel, which is then stained and viewed. However, the presence of non-viral dsRNAs in healthy plants and the apparent absence of dsRNA profile in some viruses may result in false negatives or false positives. Negative dsRNA tests should be confirmed by other methods before plant material is indexed as virus-free.

iv. Microarrays-high-throughput technology

In the context of phytodiagnostics, the simplest analogy that could be drawn is essentially a dot-blot in reverse, where the probe rather than the sample is bound to the solid phase. The logical extension of this approach is to immobilize a number of different spatially separated probes to the solid phase such that the samples can be tested for multiple targets. DNA capture probes (or spots) for each of the genes/pathogens to be detected are immobilized onto a solid support in a spatially separated and individually addressable fashion. Nucleic acid from the sample to be tested is extracted and labeled, and this labeled nucleic acid (known as the target) is then hybridized to the array. The array is scanned such that the hybridization events can be identified, and the presence of the gene/pathogen or insect pest is resolved by the pre-defined position of the DNA capture probe in the array.

Microarrays look promising for high-throughput analysis *i.e.*, screening of multiple viruses/bacteria simultaneously, provided the relevant sequence information is available. Microarray analysis in theory can combine detection, identification and quantification of a large number of fungi/bacteria/viruses/nematodes/insect pests in one single assay. Hundreds of tests could be run simultaneously and in a cost-effective manner.

v. Loop mediated isothermal amplification

Loop-mediated isothermal amplification (LAMP) is a novel technique that requires only one enzyme having strand displacement activity for amplification under isothermal conditions. LAMP has a higher specificity than PCR because its four primers recognize six distinct regions on the targeted genome. LAMP has been successfully used for detection of *High plains virus* (Arif *et al.,* 2012).

vi. Helicase dependent amplification

Helicase dependant amplification (HDA), requires no thermocycler for enhanced isothermal DNA amplification and has been successfully used for detection of *Bean pod mottle virus* (Chalam *et al.,* 2012a).

The various techniques, conventional and modern, that are employed for seed health testing of different pathogens and insect pests are summarized in Table 4.1.

Table 4.1: Summary of Various Techniques for Detecting Pathogens and Insect Pests

Techniques	Fungi	Bacteria	Viruses	Nematodes	Insects
Conventional					
Dry seed examination	+	+	+	+	+
Seed washing test	+	+	−	−	−
Soaked seed test	+	−	−	+	−
Whole embryo test	+	−	−	−	−
Incubation tests	+	+	−	−	−
Phage sensitivity test	−	+	−	−	−
Staining of inclusion bodies	−	−	+	−	−
Electron microscopy	−	−	+	−	−
Growing-on test	+	+	+	+	−
Infectivity test	+	+	+	−	−
X ray radiography	−	−	−	−	+
Transparency test	−	−	−	−	+
Serological					
Enzyme-linked immunosorbent assay (ELISA)	−	+	+	−	−
Dot-immunobinding assay (DIBA)	−	−	+	−	−
Immunosorbent electron microscopy (ISEM)	−	−	+	−	−
Lateral flow strips	−	+	+	−	−
Molecular					
Polymerase chain reaction (PCR)	+	+	+	+	+
Reverse transcription-PCR (RT-PCR)	−	−	+	−	−
Immunocapture-RT-PCR (IC-RT-PCR)	−	−	+	−	−
Real-time PCR	+	+	+	+	+
Real-time RT-PCR	−	−	+	−	−
Microarrays	+	+	+	+	+
Loop mediated isothermal amplification (LAMP)	+	+	+	+	+
Helicase dependent amplification (HDA)	+	+	+	+	+

Source: Adapted Khetarpal (2004).

For developing a standard technique, International seed testing association (ISTA), International seed health initiative (ISHI), National seed health system (NSHS) etc. involve different laboratories to carry out comparative seed health testing by using sub-samples of a given seed lot with a known level of infection and by adopting a pre-defined detection protocol. The results of comparative testing are reviewed and finally validated.

Following are the seed health testing methods developed/validated by ISTA. Some methods were sponsored by ISHI.

7-001a- *Daucus carota, Alternaria dauci*

7-001b- *Daucus carota, Alternaria dauci*

7-002a- *Daucus carota, Alternaria radicina*

7-002b- *Daucus carota, Alternaria radicina*

7-003 - *Helianthus annuus; Botrytis cinerea* (amended 02.02.2012)

7-004- Brassicaceae; *Leptosphaeria maculans*

7-005- *Pisum sativum, Ascochyta pisi*

7-006- *Phaseolus vulgaris, Colletotrichum lindemuthianum*

7-007- *Linum usitatissimum, Botrytis cinerea*

7-008- *Picea engelmannii* and *Picea glauca, Calosypha fulgens*

7-009 - *Pinus taeda* and *Pinus elliottii; Fusarium moniliforme* var. *subglutinans* (amended 02.02.2012)

7-010- *Oryza sativa, Drechslera oryzae*

7-011- *Oryza sativa, Pyricularia oryzae*

7-012- *Oryza sativa, Alternaria padwickii*

7-013a- *Hordeum vulgare; Ustilago nuda* (amended 02.02.2012)

7-013b- *Hordeum vulgare; Ustilago nuda* (amended 02.02.2012)

7-014- *Triticum aestivum; Septoria nodorum* (amended 02.02.2012)

7-015- *Festuca arundinacea, Neotyphodium coenophialum*

7-016- *Glycine max; Phomopsis complex* (amended 02.02.2012)

7-017- *Linum usitatissimum, Alternaria linicola*

7-018- *Linum usitatissimum, Colletotrichum lini*

7-019- *Brassica* spp.; *Xanthomonas campestris* pv. *campestris* (amended: 02.02.2012)

7-020- *Xanthomonas hortorum* pv. *carotae* on *Daucus carota* (amended: 02.02.2012)

7-021- *Xanthomonas axonopodis* pv. *phaseoli* and *Xanthomonas axonopodis* pv. *phaseoli* var. *fuscans* on *Phaseolus vulgaris* (amended 02.02.2012)

7-022- Agar method for the detection of *Microdochium nivale* on *Triticum* spp. (amended 02.02.2012)

7-023- Detection of *Pseudomonas savastanoi* pv. *phaseolicola* on *Phaseolus vulgaris*

7-024- Detection of *Pea early-browning virus* and *Pea seed-borne mosaic virus* on *Pisum sativum*

7-025- Detection of *Aphelenchoides besseyi* on *Oryza sativa*

7-026- Detection of *Squash mosaic virus, Cucumber green mottle mosaic virus* and *Melon necrotic spot virus* in cucurbits

7-027 - Osmotic method for the detection of *Pyrenophora teres* and *Pyrenophora graminea* on *Hordeum vulgare*.

Seed Certification

Seed certification for a crop comprises of legal norms to be qualified for ensuring genetic identity, physical purity, germinability and freedom from seed-transmitted pathogens and weeds.

ISTA and Association of Official Seed Certifying Agencies (AOSCA) among others have introduced minimum seed certification standards. Essentially, the certification procedures ensure the genetic purity and quality of seed production in the field, during harvest, processing, storage and finally inspection in the market. Seeds are distributed to farmers under the guarantee of quality, in terms of genetic and physical purity and germination capacity. Certification for seed-borne pathogens is followed only as and when required depending on the impact of the pathogen on yields.

Methodology for Quality Control of Seeds

In seed testing stations, many seed lots need to be tested and in case of viruses even very low rates of infection have to be detected in large samples. Biological assays require time for standardization, are too laborious and also time and space consuming for working with bulk samples. Testing of seeds in groups thus becomes imperative. Therefore, a practical technique has to be adopted and standardized for detecting virus in embryos. Maury and Khetarpal (1997) discussed in depth the use of ELISA for detecting viruses in single embryo, determination of seed transmission by coupling it with group analysis, mode of eliminating the interference of non-embryonic tissues (which do not play a role in transmission of virus through seeds) in routine assessment of seed transmission rate and its role in seed certification programmes. The removal of the seed testas for extraction of the antigen only from the embryos is the most tedious task while working with bulk samples. However, in case of *Pea seed-borne mosaic virus* (PSbMV)/pea Masmoudi *et al.* (1994) have obtained an antiserum specific to the N-terminal region of the viral coat protein, which is most immunogenic, and detects the virus specifically in embryos and not in testas. This was based on the observation that the capsid was partially cleaved in the testas during the seed maturation process, and such partial cleavage of the N and C terminal regions of the capsid is well known for potyviruses. This antiserum greatly simplified the extraction protocol by enabling the use of whole seeds for testing without prior decortication.

Group Testing of Seeds for Quality Control of Seed-Transmitted Viruses

A large number of seeds of a bulk seed lot is divided into a number of groups of equal size for group testing. Different groups are tested in ELISA as individual composite samples. The decision on the acceptance or rejection of a seed lot can be taken either on the basis of assessment of seed transmission or by positioning the seed lot in relation to a level of tolerance. These two alternative approaches are discussed below:

Quality Control by Assessing the Seed Transmission Rate

This consists of dividing a representative sample of the seed lot to be analyzed into N groups of n seeds. The most probable percentage of transmission, $P = 1 - (Y/N)^{1/n}$ can be estimated as a function of the number of ELISA negative groups (Y). This is demonstrated graphically for a range of seed transmission rates (Maury *et al.*, 1985). The precision of the method depends on the magnitude of the confidence interval for a given level of probability. Mathematical curves help to determine the minimum number of groups to be examined; moreover there is always a gain in precision when the number of groups (N) increases. There are other curves, which give, as a function of the percentage of transmission, the group size limit (limit for n). In such cases N and n would then be chosen as a compromise between economy and the precision required. Therefore, for routine seed health testing a workable group size limit should be determined for each host-virus system (Maury *et al.*, 1985; Maury *et al.*, 1987). It is important to verify that the percentage of ELISA-positive embryos is the same as that of infected seedlings raised from the same seed lot.

Quality Control by Positioning the Seed Lot in Relation to a Level of Tolerance

This alternative approach does not involve the determination of the percentage of transmission of the seed lots. The procedure (Geng *et al.*, 1983) takes into account a non-tolerable level of infection, decided according to its potential for causing economic losses in crop production. A (lower) tolerable level may also be defined according to general usage in quality control. Recently, Masmoudi *et al.* (1994) extended this approach to the use of group analysis for detecting PSbMV in seeds. The procedure consists of testing k groups of N seeds; the decision rule is: "the seed lot is rejected if at least one group of N seeds is found infected". The number of k groups to be tested can be determined mathematically. This approach enables the use of large group sizes. The number of tests is reduced accordingly. For example, in the PSbMV/pea seed system where inoculum thresholds of 0.1 per cent and 0.5 per cent can be considered, the number of groups to be tested may be as low as 31 and 7 groups of 200 seeds, respectively, to enable a systematic quality control of pea seed to be undertaken (Masmoudi *et al.*, 1994). This strategy is ideal for certification of seed lots.

Seed Health Certification in India

Recently due to liberalization of trade policies under WTO many developing countries have entered into seed trade. However, most of them are still not able to comply with international standards of seed health certification either due to lack of appropriate legislative measures and infrastructure or expertise in the field, which renders them less competitive. Most of them are still struggling to adopt international standards for seed health testing and to get their laboratories accredited which is mandatory for them to maintain a suitable place in the seed trade.

In India, the *Seeds Act, 1966* (including the *Seeds Bill 2004*) does not require a mandatory seed certification against any pathogens including viruses, which are the most dangerous pathogens as they cannot be controlled by ordinary physicochemical methods and require sophisticated techniques for proper detection and identification.

Besides, out of 110 crops for which seed certification standards are prescribed, seed health standards for seed-borne diseases are available only for 43 crops (59 fungal diseases, 17 bacterial diseases, 14 viral diseases and one phytoplasma disease) by seed crop inspection at field stage, for two crops (2 fungal diseases and 2 bacterial diseases) by seed sample analysis at seed stage and for seven crops by both field inspection and seed analysis. Thus, only in 9 crops including potato and sweet potato, post-harvest pathology is related to seed certification. These crops cover 16 fungal diseases, 4 bacterial diseases, one nematode disease and one bacteria + nematode complex only (Khetarpal *et al.*, 2006b).

About 130 plant viruses are known to be seed-transmitted of which one third has great economic importance, but there are no seed health standards prescribed for viral diseases at seed stage. Also, seed certification for pathogen infection during storage is not mandatory with regard to certified packed seed in store/under storage. Seed analysis is carried out essentially by dry seed examination, though many advanced detection techniques are available.

Model Case for Developing Certification Norms for Seed-Transmitted Viruses of Grain Legumes in India

Certification is an important means of managing seed-transmitted viral diseases, which are otherwise not easy to control. Keeping in view the high economic significance of seed-transmitted viruses and the complete absence of seed certification standards for them, initiatives were taken in the year 2000 at National Bureau of Plant Genetic Resources (NBPGR), New Delhi to develop a model system for seed certification for viruses in collaboration with Gujarat Agricultural University, Anand and University of Mysore, Mysore on important seed-transmitted viruses of grain legumes *viz., Bean common mosaic virus* (BCMV) and urdbean leaf crinkle disease (ULCD) of urdbean and mungbean, *Black-eye cowpea mosaic virus* (BlCMV, now a strain of BCMV) and *Cowpea aphid-borne mosaic virus* (CABMV) of cowpea, *Soybean mosaic virus* (SMV) of soybean and *Pea seed-borne mosaic virus* (PSbMV) of pea for generating information on epidemiological parameters to be used in a quality control programme and to develop a model system of seed certification for some of them. Based on extensive surveys carried out for three years in nine major legume-growing states in India complemented by testing 972 seed samples collected from diverse agencies from 21 states, a national map on prevalence of seed-transmitted viruses of grain legumes was prepared and studies revealed that the disease incidence varied with the location and the crop variety. The detection and identification of viruses both in leaves and seeds were done by deploying a combination of growing-on test, infectivity assay, electron microscopy, ELISA and RT-PCR. Studies on the samples tested by growing-on test and ELISA testing of leaf samples revealed a seed transmission rate of upto 16 per cent of BICMV and 28 per cent of CABMV in cowpea; upto 67 per cent and 49 per cent of BCMV in urdbean and mungbean, respectively; upto 55 per cent of PSbMV in pea; upto 52 per cent of SMV in soybean; upto 55 per cent and 6 per cent ULCD in urdbean and mungbean, respectively. The results gave preliminary indication on number of sites in different states that were found to be free from certain viral diseases. A correlation in viral disease incidence with aphid vector population, and appreciable losses in seed yield were observed. Based on virus spread using a known level of initial seed/seedling infection, the seed standards for certification against viruses of cowpea and soybean were proposed to be 0.5 per cent and for pea as 2 per cent. ELISA-based diagnostic kits against BlCMV and SMV were prepared to be efficiently utilized for quality control of seeds. For testing seed samples in bulk, further studies on group testing of seeds coupled with ELISA is needed on case by case basis. It is expected that the results would contribute in developing a national programme for seed certification of grain legumes (Khetarpal *et al.*, 2003; Chalam *et al.*, 2004, 2008; Chalam and Khetarpal, 2007).

National Certification System for Tissue Culture-Raised Plants

National certification system for tissue culture-raised plants (NCS-TCP) has been developed for the first time in the world by Department of Biotechnology (DBT), Government of India, where currently no such organized structure exists for certification of tissue culture material. The DBT has been authorized as the Certification Agency by Ministry of Agriculture *vide* the Gazette of India Notification dated 10th March 2006 under section 8 of the Seeds Act, 1966. Accordingly, DBT has established NCS-TCP to facilitate certification of the tissue culture raised plants up to laboratory level. NCS-TCP is the unique quality management system for tissue culture industry. This is a very comprehensive system closely associated with all the stakeholders *viz.*, Tissue culture certification agency (DBT), Accreditation unit (AU) and Project management unit (PMU) at Biotech Consortium India Limited (BCIL), Referral Centres, Accredited test laboratories (ATLs), recognized tissue culture production facilities and State Agriculture/Horticulture departments to ensure production and distribution of quality tissue culture plants.

Five test laboratories have been accredited by DBT for testing and certification of tissue culture raised plants. Till January 11, 2016 100 tissue culture production facilities have been recognized based on infrastructure, technical competency and package and practice of production of tissue culture plants. Accredited test laboratories (ATLs) will test and certify the tissue culture plants produced by recognized tissue culture production facilities.

Exclusion of Exotic Pests through Quarantine

International Scenario

The recent trade related developments in international activities and the thrust of the WTO Agreements imply that countries need to update their quarantine or plant health services to facilitate pest-free import/export. The establishment of the WTO in 1995 has provided unlimited opportunities for international trade of agricultural products. History has witnessed the devastating effects resulting from diseases and insect pests introduced along with the international movement of planting material, agricultural produce and products. It is only recently that the legal standards have come up in the form of Sanitary and Phytosanitary (SPS) Measures for regulating the international trade.

SPS measures are defined as any measure applied within the territory of the Member State:

☆ to protect animal or plant life or health from risks arising from the entry, establishment or spread of pests, diseases, disease- carrying/causing organisms;

☆ to protect human or animal life or health from risks arising from additives, contaminants, toxins or disease causing organisms in food, beverages or foodstuffs;

☆ to protect human life or health from risks arising from diseases carried by animals, plants or their products, or from the entry, establishment/ spread of pests; or

☆ to prevent or limit other damage from the entry, establishment or spread of pests.

The SPS Agreement explicitly refers to three standard-setting international organizations commonly called as the 'three sisters' whose activities are considered to be particularly relevant to its objectives: International Plant Protection Convention (IPPC) of Food and Agriculture Organization (FAO) of the United Nations, World Organization for Animal Health (OIE) and Codex Alimentarius Commission of Joint FAO/WHO. The IPPC develops the International Standards for Phytosanitary Measures (ISPMs), which provide guidelines on pest prevention, detection and eradication. To date, **thirty six standards** have been developed and several others are at different stages of development

National Scenario

Plant quarantine is a government endeavour enforced through legislative measures to regulate the introduction of planting material, plant products, soil, living organisms etc. in order to prevent inadvertent introduction of pests and pathogens harmful to the agriculture of a region and if introduced, prevent their establishment and further spread.

As early as in 1914, the Government of India passed a comprehensive Act, known as Destructive Insects and Pests (DIP) Act, to regulate or prohibit the import of any article into India likely to carry any pest that may be destructive to any crop, or from one state to another. The DIP Act has since undergone several amendments. In October 1988, New Policy on Seed Development was announced, liberalizing the import of seeds and other planting material. In view of this, Plants, Fruits and Seeds (Regulation of import into India) Order (PFS Order), first promulgated in 1984 was revised in 1989. The PFS Order was further revised in the light of World Trade Organization (WTO) Agreements and the Plant Quarantine (Regulation of Import into India) Order 2003 [hereafter referred to as PQ Order], came into force on January 1, 2004 to comply with the Sanitary and Phytosanitary Agreement (Khetarpal *et al.*, 2006a). A number of amendments of the PQ Order were notified, revising definitions, clarifying specific queries raised by quarantine authorities of various countries, with revised lists of crops under the Schedules VI, VII and quarantine weed species under Schedule VIII. The revised list under Schedules VI and VII now include 777 and 292 crops/commodities, respectively, and Schedule VIII now include 31 quarantine weed species. The PQ Order ensures the incorporation of "Additional/ Special Declarations" for import commodities free from quarantine pests, on the basis of pest risk analysis (PRA) following international norms, particularly for seed/planting material.

The Directorate of Plant Protection, Quarantine and Storage (DPPQS) under the Ministry of Agriculture and Farmers Welfare is responsible for enforcing quarantine regulations and for quarantine inspection and disinfestation of agricultural commodities. The quarantine processing of bulk consignments of grain/pulses etc. for consumption and seed/planting material for sowing are undertaken by the 57 Plant Quarantine Stations located in different parts of the country and many pests were intercepted in imported consignments. Import of bulk material for sowing/ planting purposes are authorized only through five Plant Quarantine Stations. There are 41 Inspection Authorities who inspect the consignment being grown in isolation in different parts of the country. Besides, DPPQS has developed 22 standards on various phytosanitary issues such as on PRA, pest-free areas for fruit flies and stone weevils, certification of facilities for treatment of wood packaging material, methyl bromide fumigation *etc.* Also, six Standard Operating Procedures have been notified including Export inspection and phytosanitary certification of plants/plant products and other regulated articles, Post-entry quarantine inspection etc. (www. plantquarantineindia.nic.in).

The ICAR-National Bureau of Plant Genetic Resources (NBPGR), the nodal institution for exchange of plant genetic resources (PGR) has been empowered under the PQ Order to handle quarantine processing of germplasm including transgenic planting material imported for research purposes into the country by both public and private sectors. ICAR-NBPGR has developed well-equipped laboratories and post-entry quarantine green house complex. Keeping in view the biosafety requirements, National Containment Facility of level-4 (CL-4) has been established at ICAR-NBPGR to ensure that no viable biological material/pollen/pathogen enters or leaves the facility during quarantine processing of transgenics. At ICAR-

NBPGR, adopting a workable strategy, a number of pests of great economic and quarantine importance have been intercepted on exotic material, many of which are yet not reported from India *viz.*, insects like *Acanthoscelides obtectus* in *Cajanus cajan*, *Anthonomus grandis* in *Gossypium* spp., *Ephestis elutella* in *Macadamia* nuts and *Vigna* spp., *Quadrastichodella eucalytii* in Eucalyptus, nematodes like *Heterodera schachtii*, *Ditylenchus dipsaci*, *D. destructor*, *Rhadinaphelenchus cocophilus*, etc. in soil clods and plant debris, fungi like *Claviceps purpurea* in seeds of wheat and barley, *Peronospora manshurica* on soybean, *Fusarium nivale* on wheat, barley and *Aegilops*, *Uromyces betae* on sugarbeet, bacteria like *Xanthomonas campestris* pv. *campestris* on *Brassica* spp. and viruses like *Barley stripe mosaic virus* in barley, *Broad bean stain virus* in broad bean, *Cherry leaf roll virus* on French bean and soybean, *Cowpea mottle virus* in cowpea and Bambara groundnut, *Tomato ring spot virus* on soybean etc. (Khetarpal *et al.*, 2006a; Chalam *et al.*, 2005, 2008; Chalam and Khetarpal, 2008; Chalam *et al.*, 2014). Till date, >13,000 samples of transgenic crops comprising *Arabidopsis thaliana*, *Brassica* spp., chickpea, corn, cotton, potato, rice, soybean, tobacco, tomato and wheat with different traits imported into India for research purposes were processed for quarantine clearance, wherein they are tested for associated exotic pests, if any, and also for ensuring the absence of terminator gene technology (embryogenesis deactivator gene), which are mandatory legislative requirements. A number of economically important pests (insects, mites, nematodes, fungi, bacteria and viruses) were intercepted including the ones not reported from India such as *Peronospora manshurica* (downy mildew fungus) on soybean, *Barley stripe mosaic virus* and *Wheat streak mosaic virus* on wheat. Also, *Maize dwarf mosaic virus* not reported on wheat in India was intercepted (Singh *et al.*, 2003; Chalam *et al.*, 2009).

All the plants infected by the viruses were uprooted and incinerated. The infested/infected samples were salvaged by using suitable techniques and the pest-free germplasm was only used for further distribution and conservation. If not intercepted, some of the above quarantine pests could have been introduced into our agricultural fields and caused havoc to our productions. Thus, apart from eliminating the introduction of exotic pests from our crop improvement programmes, the harvest obtained from pest-free plants ensured conservation of pest-free exotic germplasm in the National Gene Bank.

Perspectives

Detection and diagnosis of diseases and insect pests are crucial for application of mitigation strategies, seed trade and for exchange of germplasm. But, the level of confidence, knowledge and accuracy on the part of the workers need to be improved for precise detection and diagnosis. There is a need to seek technical assistance of international agencies in the area of human resource development.

Accredit diagnostic laboratories at central and state level including some well-equipped laboratories in private sector for quick and accurate identification of diseases and insect pests and there is a need for National certification programme for seed health and need to review seed certification standards proposed in Seed Bill, 2004.

Establishment of a *National repository of diagnostics for diseases and insect pests* including antisera bank, database of primers, seeds of indicator hosts, virus reference collections (lyophilized positive controls), farmer-friendly diagnostics such as lateral flow strips/dip sticks which can detect multiple viruses, multiplex RT-PCR protocols, LAMP and HDA protocols for detection of diseases and insect pests in the field and at ports of entry, microarrays, DNA barcoding and ultimately, a national biosecurity chip for diagnosis of all current threats to crop plants would be the backbone for strengthening the programme on plant health. Also *Regional working groups of experts for detection and identification of diseases and insect pests* thus need to be formed to explore cooperation in terms of sharing of expertise and facilities.

Regular survey and surveillance is required to map the distribution of diseases and insect pests and to identify hotspots. Since SAUs, State Departments of Agriculture need to play a major role in this activity, special budget outlay should be given for survey and surveillance programme. *Database on all diseases and insect pests*, including information on host range, geographical distribution, strains, etc. should be made available for its use as a ready reckoner by the scientists, extension workers and quarantine personnel. NBPGR has compiled pests of quarantine significance for cereals (Dev *et al.*, 2005) and grain legumes (Chalam *et al.*, 2012c). The Crop Protection Compendium of CAB International, UK, is a useful asset to scan for global pest data (CAB *International*, 2007).

Use simulation models for developing an *early warning system* to predict outbreaks of pests and diseases. Remote sensing may also be used for the same.

Need to develop *rapid response teams* at state level to deal with the sudden outbreak of diseases and insect pests.

Establish proper authenticity for reports of new pests/geographical distribution and deposition of reference cultures in the National Repositories may be made mandatory.

Develop the *web-based information portal* for management of plant health information and regulations including database on taxonomists and diagnosticians.

There is an urgent need to develop a *National Plant Pests Diagnostic and Certification Network* linking the research laboratories with seed/vegetative planting material testing laboratories and quarantine stations, which would be the backbone for strengthening the programme on diagnostics and certification for pests.

The NPPDCN as proposed in Figure 4.1 can be a store-house of information on diseases and insect pests, diagnostics procedures, policies, and related issues.

Finally, the quality of seed and other planting material will be further enhanced only if health of seed and other planting material is integrated into certification procedures. We have the potential for strengthening the system but need to focus in a pragmatic manner.

The techniques for rapid, specific and sensitive detection of pests have improved in terms of quality and variety during the last few years. To the extent possible,

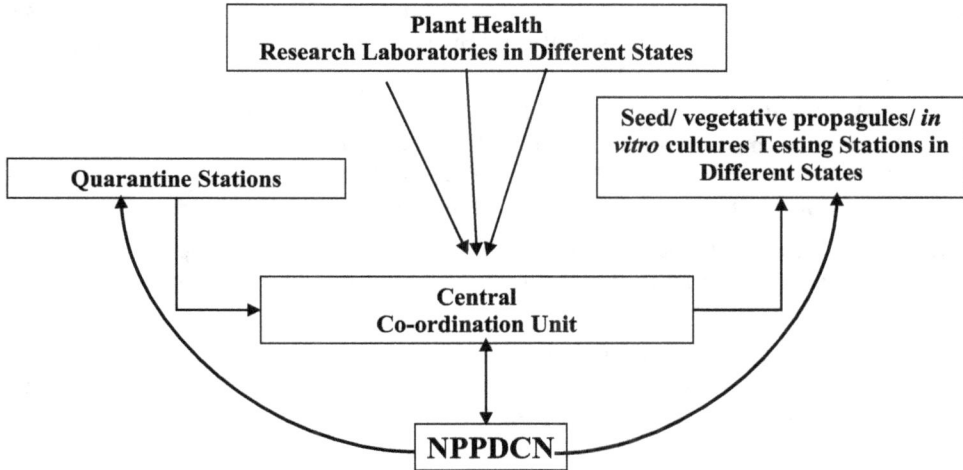

Figure 4.1: Proposed Flow Chart of Networking by National Plant Pests Diagnostic and Certification Network (NPPDCN).

the new technologies should be integrated with conventional tools now in use, so as to complement but not to substitute the latter. This will help in gaining useful information in understanding as well as controlling plant diseases and insect pests.

References

Albrechsten S E 2006. Testing Methods for Seed-transmitted Viruses: Principles and Protocols. CAB International, Wallingford, UK. 268 pp.

Arif M, Daniels J, Chalam V C, Fletcher J and Ochoa Corona F M 2012. Detection of *High plains virus* with loop-mediated isothermal amplification and qPCR. 2012 American Phytopathological Society Meeting, August 4-8, 2012, Rhode Island, USA.

Boonham N, Tomlinson J and Mumford R 2007. Microarrays for rapid identification of plant viruses. *Annual Review of Phytopathology* **45:** 307-328.

Central Seed Certification Board 1988. *Indian Minimum Seed Certification Standards,* Department of Agriculture and Cooperation, Ministry of Agriculture, Government of India, New Delhi, India. 388 pp.

CAB International 2007 Crop Protection Compendium, Wallingford, UK: CAB *International.*

Chalam V C, Arif M, Fletcher J and Ochoa-Corona 2012a. Detection of *Bean pod mottle virus* using RT-PCR, RT-qPCR and isothermal amplification. 2012 APS Meeting, August 4-8, 2012, Rhode Island, USA.

Chalam VC, M Arif, DR Caasi, J Fletcher and Ochoa-Corona F M 2012b. Discrimination among CLRV, GFLV and ToRSV using multiplex RT-PCR. 2012. APS Meeting, August 4-8, 2012, Rhode Island, USA.

Chalam V C, Bhalla S, Singh B and Rajan (eds) 2012c. Potential Quarantine Pests for India in Grain Legumes. National Bureau of Plant Genetic Resources, New Delhi, India. 324 pp. + xii.

Chalam V C and Khetarpal R K 2007. Quality control of seeds for viruses: a case study of legume viruses. In: *Seed Health Tesing and Certification: Need for Marching Ahead*, November 17, 2007, Indian Phytopathological Society and National Bureau of Plant Genetic Resources, New Delhi, India. Pp 23-28.

Chalam V C and Khetarpal R K 2008. A critical appraisal of challenges in exclusion of plant viruses during transboundary movement of seeds. *Indian Journal of Virology* **19**: 139-149.

Chalam V C D B Parakh, A K Maurya, S Singh and R K Khetarpal 2014. Biosecuring India from seed-transmitted viruses. The case of quarantine monitoring of legume germplasm exported during 2001-2010. *Indian Journal of Plant Protection* **42**: 270-279.

Chalam V C, Khetarpal R K, Parakh D B, Maurya A K, Jain A and Singh S 2005 Interception of seed-transmitted viruses in French bean germplasm imported during 2002-03. *Indian Journal of Plant Protection* **33**: 134-138.

Chalam V C, Khetarpal R K, Prakash H S and Mishra A 2008. Quality control of seeds for management of seed-transmitted viral diseases of grain legumes in India. In: Kharakwal M C (ed.) Food Legumes for Nutritional Security and Sustainable Agriculture, Vol. 2, Proceedings of the Fourth *International Food Legumes Research Conference* (IFLRC-IV), Indian Society of Genetics and Plant Breeding, New Delhi, India. Pp 468-474.

Chalam V C, Parakh D B, Khetarpal R K, Maurya A K and Digvender Pal 2009. Interception of seed-transmitted viruses in broad bean (*Vicia faba* L.) germplasm imported into India during 1996-2006. *Indian Journal of Virology* **20**: 83-87.

Chalam V C, Parakh D B, Khetarpal R K, Maurya A K, Jain A and Singh S 2008. Interception of seed-transmitted viruses in cowpea and mungbean germplasm imported during 2003. *Indian Journal of Virology* **19**: 12-16.

Dev U, Khetarpal R K, Agarwal P C, Lal A, Kapur M L, Gupta K and Parakh D B (eds.) 2005. Pests of Quarantine Significance in Cereals. National Bureau of Plant Genetic Resources, New Delhi, India. 142 pp.

Geng S, Campbell R N, Carter M and Hills F J 1983. Quality control programs for seed borne pathogens. *Plant Disease* **67**: 236-242.

http://seedtest.org/en/download-ista-seed-health-testing-methods-_content—1—1132—746.html

https://www.ippc.int/index.php?id=13399 and type=publication and subtype= and category_id= and tx_publication_pi1[pointer]=0 and showAll=1#publication.

http://www.plantquarantineindia.nic.in/.

Khetarpal R K 2004. A critical appraisal of seed health certification and transboundary movement of seeds under WTO regime. *Indian Phytopathology* **57**: 408-421.

Khetarpal R K, Lal A, Varaprasad K S, Agarwal P C, Bhalla S, Chalam V C and Gupta K 2006a. Quarantine for safe exchange of plant genetic resources. In: AK Singh, Kalyani Srinivasan, Sanjeev Saxena and BS Dhillon (eds.) *Hundred Years of Plant Genetic Resources Management in India*, National Bureau of Plant Genetic Resources, New Delhi, India. Pp 83-108.

Khetarpal R K, Sankaran V, Chalam V C and Gupta K 2006b. Seed health testing for certification and SPS/WTO requirements. pp 239-258 In: Kalloo G, Jain S K, Alice K Vani and Srivastava U (eds.) *Seed: A Global Perspective*, Indian Society of Seed Technology, New Delhi, India. 312 pp.

Mathur S B and Olga Kongsdal 2003. Common Laboratory Seed Health Testing Methods for Detecting Fungi. International Seed Testing Association, Switzerland. 425 pp.

Masmoudi K, Duby C, Suhas M, Guo J Q, Guyot L, Olivier V, Taylor J and Maury Y 1994. Quality control of pea seed for pea seed borne mosaic virus. *Seed Science and Technology* 22: 407-414.

Maury Y, Bossennec J M, Boudazin G, Hampton R O, Pietersen G and Maguire J D 1987. Factors influencing ELISA evaluation of transmission of pea seed borne mosaic virus in infected pea seed: seed group size and seed decortication. *Agronomie* 7: 225-230.

Maury Y, Duby C and Khetarpal R K 1998. Seed certification for viruses. pp 237-248 In: Hadidi A, Khetarpal R K and Koganezawa H (eds.) *Plant Virus Disease Control*, American Phytopathological Society (APS) Press, St. Paul, Minnesota, USA. 684 pp.

Maury Y and Khetarpal R K 1997. Quality control of seed for viruses: Present status and future prospects. In: J D Hutchins and Reeves J E (eds.) Seed Health Testing: Progress towards the 21st Century, CAB International, U.K. Pp 243-252.

Mumford R, Boonmah N, Tomlinson J and Barker I 2006. Advances in molecular phytodiagnostics – new solutions for old problems. *European Journal of Plant Pathology* 116: 1-19.

Chapter 5

Vision and Action Plan for Managing Emerging Pests towards Ensuring Food Security

**G V Ranga Rao[1]*, V Rameshwar Rao[1]
and T Ramesh Babu[2]**

[1]*International Crops Research Institute for the Semi-Arid Tropics (ICRISAT),
Patancheru – 502 324, Telangana State*
[2]*Prof Jaya Shankar Telangana State Agricultural University(PJTSAU),
Rajendranagar, Hyderabad – 500 030, Telangana State*
*E-mail: g.rangarao@cgiar.org

ABSTRACT

Insect pests are the integral part of agriculture where they play an important role in ensuring food security. These populations are dynamic in nature and often rapidly grow and cause economic concerns. Such sporadic outbreaks can be catastrophic and result in complete destruction of crops. These disturbances in developing countries are very common, resulting in national economic imbalance, threatening stability of food security. Pest situations in the past four decades revealed several changes in their populations and shifts in their economic importance such as the grand old pest of polyphagus nature, red hairy caterpillar, Amsacta albistriga, white grub species, Holotrichia serrata, groundnut leafminer, Aproaerema modicella in the southern parts of India lost their prime stature and are no longer a threat in wider areas. Some species such as coconut eriophyid mite, Aceria guerreronis (Keifer), rice panicle mite, Steneotarsonemus spinki Smiley, armyworm, Spodoptera exigua (Hubner), legume borer, Maruca vitrata (Geyer), mealybug, Phenacoccus solenopsis Tinsley, mango fruit borer, Deanolis albizonalis (Hampson), bud mite, Aceria mangiferae Sayed, thrips as vectors of viral diseases in groundnut, and termites, which were not considered as economic in 70s gained prominent place in the present day plant protection. The importance of internal feeders such as stem borers, boll worm/legume pod borer Helicoverpa armigera (Hub.) and tobacco caterpillar, Spodoptera litura (Fab.) mostly maintained their status. The invasive species such as groundnut leafminer in east Africa is of much concern. In general, the virus vectors (Thrips, aphids and mites) and storage pests, pulse bruchid, Callasobruchus chinensis *(Fabricius) and groundnut bruchid* Caryedon serratus Oliver *have gained significant importance. Despite the significant adverse impacts and indications of several out breaks, there has been no attempt to gather and systematically analyse the information on the changing scenario of various pest species. It was felt that the investments on plant protection should be given high priority to disseminate the potential options to the fields with farmer participatory approach with high emphasis on operational and environmental safety. This paper provides status of the pest situation in key crops and their shifts with reference to Asia and Africa to update the plant protection needs with potential action plan to ensure food security.*

Keywords: *Pest dynamics, Insect pests, Pest management, Emerging pests, Food security.*

Introduction

Increasing human populations and food demand are placing unprecedented pressure on agriculture and natural resources. Today approximately one billion people are chronically undernourished while our agricultural systems are degrading land, water and biodiversity. The population pressure, coupled with changes in agricultural policies, dietary habits in developing countries towards high quality food and the increasing use of grains for livestock feed, is projected to cause the high demand for agricultural productivity. The land suitable for agricultural production is limited, and most of the soils with high productivity potential have already reached a plateau in most of the Asian countries. Safeguarding crop productivity by protecting crops from damage by weeds, insect pests and pathogens is a major requisite for providing food and feed in sufficient quantity and quality with the same natural resources (United Nations, 1996; Tilman, 1999; Pinstrup- Andersen, 2000, Foley *et al.*, 2011).

Insects and pathogens are part of all agricultural systems and when present at relatively low densities, causing little damage, and having negligible impact on crop growth and vigour under natural competition. However, as the pressure to enhance productivity increase, some species outgrow rapidly resulting in outbreaks, which may persist for a variable length of time before subsiding. Such large populations may have adverse effects on crops and affect the livelihoods of farming communities.

Crop losses caused by insect pests in India alone were estimated around 18 per cent, which is valued at Rs 90,000-crore (22 million US dollars) each year. The Indian pesticide industry with an annual production of 85,000 MT during 2007 was ranked second in Asia, only behind China and 12th globally (Anonymous, 2011). In the present day plant protection scenario, there are several concepts in practice starting from total chemical dependence to complete organic. These are two extremes which were found successful in different situations. Considering the food crisis in developing countries and the need to reduce the pre and post harvest losses from insect pests to save food for the people one has to draw a comprehensive strategy between these two concepts.

Improved crop management systems based upon improved (high-yielding) cultivars, enhanced agronomic practices, pest control *via* synthetic pesticides and integrated approaches proved effective in addressing the food crises in many Asian countries (Budy P Resosudarmo, 2001). Several changes in the cultivation practices, disturbance in the traditional cropping systems, injudicious use of chemicals resulted in the significant change in the pest dynamics in the past few decades. Several major pests of the region disappeared or attained minor importance and on the other hand less-known/unknown species attracted major attention (Amit Sethi *et al.*, 2002, Shivankar *et al.*, 2007). In view of this, it was felt necessary to update the economic importance and shifts in populations of key pests to develop effective management strategies for ensuring food security.

Present Status

As the importance of agricultural crops changing from subsistence to intensive market oriented mode, there were drastic changes occurred in different cropping systems in the past few decades. Several waste lands were brought under cultivation, on the other hand area under irrigation increased to meet the food demand. In this process several important pest species lost their diapausing sites, alternate hosts which affected their survival and status of economic importance. Intensification in food production also led to increased use of farm inputs particularly plant protection chemicals causing imbalance in pest and natural enemy ratio resulted in secondary pest out breaks. Crop intensification also lead to decline in biodiversity, increased the need for external inputs (Médiène *et al.*, 2011). One such example was the flare up of whiteflies in cotton after the introduction of pyrethroids during 1985. The excessive use of chemical also led to the development of resistance in key species such as *Helicoverpa* and *Spodoptera* to a range of chemicals in India (Armes *et al.*, 1997, Kranthi *et al.*, 2002). The trends in pest status in the past few decades clearly brought out significant shifts in several insect species across the crops (Table 5.1).

Cereals

Among various cereal pests in the past four decades, the importance of rice stem borer, *Scirpophaga incertulus* Walker and *Chilo partellus* in sorghum has reduced considerably. The species such as gall midge, which was recognized as a serious one in 70s, now it's importance has declined. The brown planthopper, *Nilaparvata lugens* Stal., which was of minor importance in 70s attained serious status in rice all over Asia. Panicle sterility mite in rice and *Helicoverpa* causing sorghum grain damage has gained considerable importance particularly in seed production.

Legumes

The pest problems in legumes increased as their economic importance raised in contributing the livelihood of the farmers in Asia. The notorious pod borer *Helicoverpa* is the one pest, which retained its number one position over the past four decades. With the introduction of different high yielding varieties to suit various cropping systems in various legume crops, particularly pigeonpea, the less known pest, *Maruca vitrata* has gained its importance all over Asia causing significant damage to pigeonpea, mung and urd beans. The incidence of pod fly *Melanagromyza obtusa,* though regular in many areas in cooler regions, now has spread its distribution to warmer areas which needs special attention. Pod sucking bugs in legumes which were never considered as serious constraint in the past have slowly gained their importance as a seed quality constraint in several legumes,

Table 5.1: Shift in the Economic Status of Insect Pests of Major Crops in India during Past Four Decades

Crop	Common Name	Scientific Name	Shift in Economic Status		
			1970	1990	2010
Paddy	Stem borer	Scirpophaga incertulus	+++	++	++
	Brown plant hopper	Nilaparvata lugens	–	++	+++
	Gall midge	Orseolia oryzae	+++	+	–
	Leaf folder	Cnaphalocrocis medinalis	–	+++	+
	Panicle mite	Steneotarsonemus spinki	–	–	++
Sorghum	Aphid	Schizaphis graminum	+++	++	+
	Stem borer	Chilo partellus	+++	+++	++
	Shoot fly	Atherigona spp.	+++	+++	++
	Earworm	Helicoverpa armigera	–	–	+
Pigeonpea	Pod borer	Helicoverpa armigera	+++	+++	+++
	Pod fly	Melanagromyza obtusa	++	++	++
	Leaf webber	Maruca vitrata	–	+	+++
	Pod sucking bugs	Clavigralla gibbosa	–	+	++
Chickpea	Pod borer	Helicoverpa armigera	+++	+++	+++
	Cutworm	Agrotis ipsilon	–	++	++
Soybean	Stem fly	Ophiomyia phasioli	–	+++	++
	Girdle beetle	Obereopsis brevis	–	–	++
Groundnut	Hairy caterpillar	Spilosoma obliqua	+++	+++	++
	Leaf miner	Aproaerema midicella	+++	++	+
	Tobacco caterpillar	Spodoptera litura	+++	++	+
	Thrips	Scirtothrips dorsalis	–	+++	+++

Contd...

Table 5.1–Contd...

Crop	Common Name	Scientific Name	Shift in Economic Status		
			1970	1990	2010
Sunflower	Aphids	*Aphis craccivora*	+++	+	−
	Termite	*Odontotermis* sp.	−	−	+
Sunflower	Gram pod borer	*Helicoverpa armigera*	+++	+++	++
Sesame	Leaf webber	*Antigastra catalaunalis*	−	++	+
Rapeseed	Aphids	*Lipaphis erysimi*	−	++	++
Brinjal	Fruit and stem borer	*Leucinodes orbanalis*	+++	+++	+++
Cabbage and Cauliflower	Diamond back moth	*Plutella xylostella*	−	+++	++
	Tobacco caterpillar	*Spodoptera litura*	+++	+++	+
Tomato	Fruit worm	*Helicoverpa armigera*	+++	++	++
Apple	San Jose scale	*Quadraspidiotus perniciosus*	+++	++	+
	Codling moth	*Cydia pomonella*	+++	+++	++
	Phytophagous mites	*Panonychus ulmi*	−	++	++
Grapes	Flea beetle	*Scelodonta stricollis*	++	+	+
	Thrips	*Retithrips syriacus*	−	+++	++
	Mealy bugs	*Maconellicoccus hirstutus*	−	−	+++
Oranges	Fruitflies	*Carpomyia vesuviana*	+	++	++
	Defoliators	*Papilio demoleus*	+++	++	+
Mango	Hopper	*Amritodes atkinsoni*	+++	+++	+++
	Leaf webber	*Orthaga exvinacea*	−	+++	+
	Stemborer	*Batocera rufomaculatus*	−	++	+
	Fruit borer	*Deanolis albizonalis*	−	+	+++
	Mites	*Aceria mangiferae*	−	−	+++

Contd...

Table 5.1–Contd...

Crop	Common Name	Scientific Name	Shift in Economic Status		
			1970	1990	2010
Cotton	American bollworm	*Helicoverpa armigera*	+++	+++	++
	Pink bollworm	*Pectinophora gossipiella*	+	+++	++
	Whitefly	*Bemisia tabaci*	−	+++	++
	Spotted bollworm	*Earias insulana*	+++	+++	+
	Mealy bug	*Phenacoccus solenopsis*	−	−	+++
Sugarcane	Stem borer	*Chilo sacchariphagus indicus*	+++	++	+
	Wooly aphid	*Ceratovacuna lanigera*	+	+	++
	Yellow mite	*Oligonychus sacchari*	+	+	++
	Pink mealy bug	*Saccharicoccus sacchari*	+	+	++
Coconut	Rhinoceros beetle	*Oryctes rhinoceros*	+++	++	+
	Red palm weevil	*Rhyncophorus ferrugineus*	+++	++	+
	Eriophyid mite	*Aceria guerreronis*	−	+	+++
	Black headed caterpillar	*Brachimeria nosatoi*	+++	++	+
	Scale insect	*Melanapsis glomerata*	+	+	+
Tobacco	Tobacco caterpillar	*Spodoptera litura*	+++	+++	++
	Whiteflies	*Bemisia tabaci*	+	++	+
Storage pests					
Cereals	Rice weevil	*Sitophilus oryzae*	+	+	+
	Paddy moth	*Sitotroga cerealella*	++	++	++
	Rice moth	*Corcyra cephalonica*	++	++	++
	Red flour beetle	*Tribolium castaneum*	+	+	+
Pulses	Bruchids	*Callasobruchus chinensis*	++	++	+++
Groundnut	Groundnut bruchid	*Caryedon serratus*	+	++	+++

including pigeonpea, cowpea, mung and urd beans. *Clavigralla* sp attained number one position in case of pigeonpea in Kenya. In chickpea, historically the pod borer, *Helicoverpa* remained important, however, in recent years, the cutworm, *Agrotis ipsilon* and *Spodoptera exigua* gained considerable importance in Asia, Stem fly and girdle beetle are gaining importance in soybean and clusterbean in Asia.

Fruits and Vegetables

Vegetable pests though are important from the beginning, their economic status has increased as the demand for vegetables increased across the world. Among various vegetable pests, brinjal fruit and stem borer only remained important across Asia and demands high levels of protection. The economic status of tomato fruit borers, *Helicoverpa* and *Spodoptera* fluctuated over time based on the importance of the crop as well as adoption of IPM strategies in the region. The damage caused by diamond back moth in cabbage and cauliflower, which was negligible in 70s gained prominence by 90s, remained as classic example of injudicious use of insecticides.

In Orchards the importance of traditional San Jose Scale and codling moth in apple have decreased while the sucking pests such as mites gained considerable status. In grapes the importance of mealybugs attracted much attention while the damage by thrips in affecting quality of fruit fluctuated over the decades. In mango, the hopper, *Amritoides atkinsoni* is the only pest that showed consistent trends over decades and still occupied the serious status. The other two important species such as leaf webber and stem borer, which were considered as minor in 70s had gained importance in 90s and currently treated as minor importance. One good example to remember in the recent past is mango fruit borer, *Deanolis albizonalis*, which was considered as minor in 90s and nobody paid attention for the introduced pest, now it became a billion dollar problem threatening the mango cultivation in Andhra Pradesh where the plant protection experts had no viable solution to face this challenge. The basic information on its developmental biology, alternate hosts, survival in off season, effective chemicals is still lacking to address such devastating pests.

Oilseeds

Insect pest situation in groundnut has taken greater turn and overall the importance of several pests has gone down particularly the defoliators such as leafminer, tobacco caterpillar and red hairy caterpillars in Asia, which were on the prime list in 70s, are now of minor importance. Groundnut leafminer, which was considered as species of Asian origin and restricted to Asia in distribution, now reported from Uganda in 1998 and within a short period spread over several countries in the east Africa (Mozambique, Malawi, Congo and South Africa) (Kenis, and Cugala, 2006). Among various sucking pests, the thrips gained importance as vector of viral diseases. The importance of aphids, which was of prime importance in the past, has reduced drastically.

The red hairy caterpillars, which were considered as the pest of several rainy season crops in 70s, slowly, lost its economic importance mainly because of significant changes in cropping systems and cultivation practices, which destroyed their hibernation sites and adoption of mass trapping of adults in endemic areas using light traps. At present their importance is confined to very limited tracts in the southern India.

In sunflower, historically the *Helicoverpa* was considered as important, causing damage to seeds, however, its economic importance had been declined due to the development and adoption of IPM in several countries.

The cultivation of mustard gained importance in rain-fed in post rainy season in Asia, as the importance of edible oil increased in the region, where aphid gained considerable importance.

Among soil insects, white grubs were of prime importance in several locations in India, Vietnam and Nepal in 90's however, at present due to drastic changes in the cropping system, cultivation practices and community actions they lost their economic status. Termite *Odontotermis* gained its economic importance in several crops including groundnut, particularly in rain-fed situations of India, Nepal and Myanmar.

Commercial Crops

Among various commercial crops, cotton registered as the important one which, required about 50 per cent of the total pesticide required in most of the countries in Asia. Among various insect pests, the boll worm complex was treated as prime importance in the past four decades. With the introduction of transgenic varieties to resist boll worms the chemical pesticide inputs on cotton were reduced significantly, however, the increased importance of sucking pests such as aphids, thrips, whiteflies were at increasing trend. Another important sucking pest (mealybug), which was not considered in economic terms until 90s has gained significance in the past decade.

The traditional coconut pests such as rhinoceros beetle, red palm weevil and black headed caterpillar slowly lost their economic status over the decades and at present are of minor importance. However, the dreaded eriophyid mite, *Aceria guerreronis* invaded Indian palm grooves probably from Northern Sri Lanka and caused serious havoc in shaking the coconut industry in India. This also caused major concern in finding potential eco-friendly solutions and application strategies against this pest.

The importance of stem borer in sugarcane and defoliators in tobacco has reduced over the years, which could be due to the adoption of natural control process in sugarcane and introduction of new molecules in case of tobacco caterpillar. The wooly aphid, *Ceratovacuna lanigera*, yellow mite, *Oligonychus sacchari* and pink mealy bug, *Saccharicoccus sacchari* are also on increase in sugarcane.

Storage

In the past, storage insect pest menace was never considered as economically important in real farm situation, except for saving seed material. But in recent years, as the agriculture moved from subsistence to market oriented approach, the importance of storage has gained considerable importance. Among various storage pests, paddy moth, *Sitotoroga cerealella* and rice moth, *Corcyra cephalonica* were considered as important throughout Asia over these decades and their status remained static. The problem of storage in legumes is recognized as important in the past decade, particularly against pulse and groundnut bruchids as the distribution of these seed-lots increased in various seed networks of the country.

Up-coming Pests

Considering the importance and the trends in the economic status of various species, *Maruca vitrata* on legumes, mealybugs in fruits, vegetables and cotton, mango fruit borer, mites and termites on number of crops particularly in dry tracks cause significant damage to the crops and their management has also led to environmental and operational risks. The pulse and groundnut bruchids are of prime importance in Asia in both seed as well as commercial stocks. The basic research on the above insect pests covering their population dynamics, distribution and viable management need to be generated, evaluated and streamlined to meet the ongoing challenges.

Deanolis albizonalis (Hampson), the red banded mango caterpillar (RMBC), is a Southeast Asian insect species. It is now widely distributed throughout this region (India, Burma, Thailand, China, Brunei, Philippines, Indonesia, and Papua New Guinea) and was recently detected for the first time on mainland of Australia. Although infestation levels of 40-50 per cent were recorded very little is known about the biology of this species, (Waterhouse 1998; Stefan Krull and Thies Basedow, 2006). The recent experience in Andhra Pradesh, India revealed the devastating status of this pest threatening the mango farmers to look for some alternative crops.

The legume pod borer, *M. vitrata* is a serious pest of grain legumes in the tropics and subtropics because of its wider host range, destructiveness and distribution. It is a serious pest of cowpea, pigeonpea, urd bean, mung bean, beans and soya beans in Asia and Africa. In the absence of host plants in the off-season the populations of *M. vitrata* survive on alternative plants like wild leguminous shrubs and trees. There is a considerable shift in its' importance in India, Vietnam, China, SriLanka and Myanmar due to its feeding habit on various legumes and the ineffectiveness of number chemicals.

Mealybug was considered to be a minor pest of cotton in the past but it emerged as a major pest in 2006-2007 in North and Central zones of India. Extensive field surveys conducted in cotton fields during 2007-09 in Haryana, Rajasthan and Punjab in the North zone and Madhya Pradesh, Maharashtra and Gujarat in the Central zone indicated that *Phenacoccus solenopsis* was the only major species of mealybug recorded on cotton in North as well as Central zones. Infestation of mealybug at most of the places in North and Central zones ranged from mild (10-20 per cent) to high (40-60 per cent) during 2007 and 2008. Considering the feeding habit and protective covering above the pest colonies special attention is required to keep this species under manageable level. In recent years it gained importance in number of fruit crops and ornamentals where the productivity and quality were of high concern.

Mites are common on many crops with particular reference to fruits and vegetables. These mites are small and often difficult to see with the unaided eye. Their colour ranges from red and brown to yellow and green, depending on the species of spider mite and seasonal changes in their appearance. Damaged areas typically appear marked with many small, light flecks, giving the plant somewhat sick appearance. Under severe infestations, leaves become discoloured, scorched and drop prematurely. Spider mites frequently kill plants or cause serious stress to them.

Eriophyid mites (*Eriophyes* spp.) are so small and difficult to see without some magnifying device that they often go undetected. The symptoms they produce include pale colour patches on leaf surfaces, leaf margins that roll inward or downward, swollen and distorted leaves, galls. Besides the direct damage, they also act as vectors of viral diseases such as sterility mosaic in pigeonpea.

Bruchids are a major problem in stored legumes in all regions and are most often not detected until seed has been stored for a reasonable period (*e.g.* for longer than three months). Bruchids breed rapidly in storage and by the time they are detected, the infested grain is usually unmarketable. Historically, bruchid infestations have been worse in seed lots at farm level, however, this damage is underestimated in most of the regions. Though farmers of Asia are familiar with some indigenous methods to save their seeds they often fail in their approach due to their ignorance. The bruchid responsible for most infestations in legumes in Asia is *Callosobruchus maculatus* (Fabricus).

The most commonly reported stored pests of groundnut in Asia and Africa is *Caryedon serratus* (Olivier) infestation in groundnut is well known for causing direct loss, but indirect loss in terms of quality of the produce also impacts its trade and use. Knowing its occurrence, identification of initial damage by groundnut bruchid particularly while exporting the nuts plays a critical role in limiting this species to save the quality of groundnut seeds.

Approach

Interaction with farming communities during 2005-07 brought out the levels of plant protection inputs in different crops ranging from 6-44 per cent involving 1-15 sprays. In spite of chemical sprays, farmers also experienced 11-40 per cent crop losses caused by insect pests (Table 5.2). The impact of ICRISAT's IPM research organized in collaboration with National agricultural research and extension systems (NARES) and non-governmental organizations (NGOs) in India showed encouraging results with drastic reduction in chemical use without sacrificing the productivity in several locations (Ranga Rao and Rameshwar Rao, 2010). Programmes on capacity building of both the extension workers and farmers in the Integrated pest management (IPM) were started throughout Asia involving national and international organizations (Anon-FAO) in rice during year 1980 for the first time with the assistance from FAO. In fact, the Government of India had adopted IPM as a cardinal principle of plant protection in 1985. Despite techno-economic superiority of IPM over conventional chemical control, adoption of IPM remains restricted to hardly two per cent of the area treated (Pratap S. Birthal and Sharma, 2004). On the other hand studies organized in a consortium approach clearly brought out the successful implementation of the concept with substantial reduction in plant protection inputs across several locations. Several IPM success stories to support consortium approach such as Ashta in Maharastra (a collaboration among

Table 5.2: Data on Crop Loss Estimates, Cost of Plant Protection, Production, and Yields Collected on some Major Crops in Andhra Pradesh, India, during 2005-2007

Crop	No. of Farmers	Crop Loss by Pests (Per cent)	No. of Chemical Sprays	Cost of Production (Rs. ha⁻¹)	Cost of Plant Protection (per cent)	Yield(t ha⁻¹)
Cotton	692	35	8	18674	40	1.8
Cotton-Bt	66	26	7	21068	33	2.3
Chilli	188	28	15	37201	44	4.1
Rice	142	19	3	12206	22	5.1
Pigeonpea	425	31	2	6514	25	0.8
Chickpea	267	29	2	7980	12	1.3
Maize	72	11	1	6901	14	3.7
Groundnut	119	40	2	8045	18	1.2
Sunflower	21	32	2	8565	12	1.1
Wheat	90	–	1	8290	7	–

ICRISAT, NCIPM, Cotton research institute, Nanded and farmers) and Punukula in Andhra Pradesh (collaboration among ICRISAT and NGOs' CWS, SECURE) were the classic examples with 22-100 per cent reduction in pesticides in a span of 2-7 years (Table 5.3).

Table 5.3: Details of Pesticide Reduction in IPM Fields Compared to Farmer Practice (1997-2000) at different Locations

Village (NGO)	Reduction in Pesticide Use (per cent)
Hamsanpalli (REEDS)	21.5
Bollibaithanda (REEDS)	36.1
Chincholi (CEAD)	46.9
Kanjar (CEAD)	55.8
Maddur (CHRD)	67.0
Panyala (ROAD)	60.9
Marlabeed (SEVA)	84.1
Punukula (SECURE)	55.0*
Deverajugattu (CAFORD)	79.1
Itagi (PRERANA)	41.6
Jeedigaddathanda (VIKASAM)	76.8
Pastapur (DDS)	28.6
Bhavanandapur (TREES)	53.5
Pothinenipalli (PILUPU)	54.3
Ashta (NCIPM/MAU)	**
Nellipaka (FRSF)	60.0

* By year 2004 this village became 100 per cent chemical pesticide free; **Whole village adopted IPM.

Though, a majority of the farmers are aware of the benefits of collective action, a number of socio-economic and technical factors that act as detrimental for the rapid spread of these programs. At this juncture one cannot expect high levels of IPM adoption when majority of the farmers initiate chemical sprays with the first occurrence of the pests rather than following their population fluctuations across the season/different crop stages. In recent years, though several plant protection chemicals were banned in the developed countries they are still popular in the developing countries mainly due to ignorance and the prevailing government policies. Some products for example, DDT and BHC, which were banned in agriculture and permitted for use in health programs for mosquito control, are widely proliferating in to agriculture. Further, many pesticides that have been banned are available to farming communities in the developing world. This is primarily happening due to the ignorance of the poor countries in Asia and Africa and lack of policy and economic strength of these governments. This is not fair and need to be addressed in every discussion until further feasible action is taken by every affected nation. Strict enforcement of the regulations governing production, marketing, distribution, use and quality of pesticides would help in their efficient use in addressing productivity and sustainability.

At present though some farmers in Asia were aware of importance of IPM and its impact on health and environment the adoption level was not up to the expected levels. However, latest estimates are quite encouraging with reduction in chemical use to $25.3 billion in 2010 compared to $26.7 billion in 2005. On the other hand interestingly the bio-pesticides market is growing rapidly from $672 million in 2005 to over $1 billion in 2010. Bio-pesticides currently has 2.5 per cent of the overall pesticides market, but its share of the market was predicted to increase to over 4.2 per cent by 2010 (Anon, 2009).

In recent years the shift from subsistence to commercial mode of farming lead to the intensive monoculture systems using high yielding varieties, resulted in an environment conducive to pest build up and infestation, and the consequent use of pesticides disrupted the natural pest-predator balance. Today, with improved varieties with resistance to insect pests, crop management strategies, and advances in integrated pest management, along with a reformed policy environment try to discourage pesticide use. Certainly in the case of insecticides for rice and fungicides for wheat, recent evidence indicates that the productivity benefits of applying chemicals are marginal. However, commercialization of these production systems has led to increased inputs include pesticides.

In addition, because early pesticide formulations were often non-selective, pesticides proved equally lethal to beneficial insects that preyed on crop pests (Pagiola, 1995; Rola and Pingali, 1993). In case of rice in tropical Asia, a large number of pest outbreaks have been associated more with injudicious pesticide applications, high cropping intensity, and/or high chemical fertilizer use (Heinrichs and Mochida, 1984; Kenmore *et al.*, 1984; Joshi *et al.*, 1992, Schoenly *et al.*, 1996). Outbreaks of secondary pests of rice, notably the brown planthopper, previously of minor significance, began to occur in regions adopting modern varieties and concomitant use of agrochemicals (Pingali and Gerpacio, 1997). High and injudicious pesticide applications disrupt the rich diversity of pest and predator populations, where in most instances the species richness and abundance of predator populations may be greater than those of the pest populations.

Progress in crop improvement is limited by the ability to identify favourable combinations of genotypes (G) and management practices (M) given the resources available in the target population of environments (E). Crop improvement can be viewed as a search strategy on a complex G × M × E adaptation or fitness landscape (Hammer *et al.*, 2007; Uaboi-Egbenni *et al.*, 2010).

Global crop production is presently sufficient to feed the human population, however hunger and malnourishment prevail in some regions because of low productivity and inefficient distribution. The increased threat of higher crop losses to pests has to be counteracted by improved crop protection by whatever method, *e.g.* biological, mechanical, chemical, or training of farmers and advisors in integrated pest management (IPM). An intensification of crop production without an adequate protection from pests will not result in any economic impact. Therefore, although a drastic reduction of crop losses is highly desirable for many regions to meet the demand to feed the growing populations. The concept of judicious use of pesticides should, on the other hand, results in lower pesticide use when crop prices are falling. All appropriate strategies are used such as enhancing natural enemies, resistant crops, adapting cultural management, and using pesticides judiciously'. Programs have been developed to reduce the dependency of production on synthetic pesticides, to minimise the effect on the environment and to maintain the efficacy of crop protection products in order to enable sustainable crop production at higher intensities. IPM programs have been established in various crops around the world and have proven their suitability in developed and developing countries (Fernandez-Cornejo, 1998; Cuyno *et al.*, 2001). IPM is successfully practised in perennial and annual crops in temperate and tropical conditions for the control of all pest groups, especially insect pests and fungal pathogens (Berg, 2001; MacHardy, 2000; Verreet *et al.*, 2000; Way and van Emden, 2000). However adoption of IPM in Asian and African continents is not up to anybodies expectation which could be due to lack of efficient Extension, knowledge, and the availability of quality inputs at farm level and shifts in the population dynamics of key species coupled with low investments.

Evaluations of farmers using IPM programs have generally found an insignificant effect on yield, a small increase in profit, and a reduction in environmental risk associated with lower use rates or improved timing of application (Brethour and Weersink, 2003). In intensive rice production in Southeast Asian countries like Indonesia, governments have implemented national and international pesticide regulation programs; in addition to strict regulation of pesticide registration and the introduction of IPM systems has contributed to a significant reduction of the pesticide use per unit of area without affecting crop productivity. In many developing countries commercial crops like cotton and vegetables take more than 50 per cent of the total pesticide used. Hence care must be taken to concentrate on these crops to capture the immediate impact.

With recent advances in developing eco-friendly approaches in insects and diseases management strategies, the trends in future reductions in toxic pesticide use are very promising.

Under low pest infestation levels natural control relies on natural enemy populations to manage pest infestations under normal circumstances. Under the present day IPM approach, pesticides should be used as a last resort. In this regard, farmers who are well versed with pest and its ecosystem can take advantage of IPM techniques. Therefore, continued investments in IPM training and adoption are essential for the successful prediction and management of epidemics.

The success of pest management programs depends on several things such as collective organization against migratory pest infestation, many times the actions of individual farmers in managing their pest problems could have detrimental effects on the community as a whole. In this regard, pest management could be viewed as a common problem and dealt with thorough effective collective action. In Asia, where farms are uniformly small and farmers nearly homogeneous, collective action for pest control seems to be quite an attractive option. However, it is not an easy task to bring all farmers under one umbrella in Asia unless a cooperative model like Vietnam is adopted. This is mainly due to the resource poor situation of the farming communities of subsistence nature.

Way Forward

In order to overcome the existing evils (Resistance, resurgence and residues) of plant protection it is believed that the following action plan is suitable for ensuring future food security with better environment.

☆ **Investment** in the development and implementation of plant protection research needs to be enhanced to arrest further degradation of natural resources due to toxic residues and to reclaim them.

☆ **Generating and sharing** data on toxic residues in food, feed and water bodies is of high priority with farmer participatory approach

☆ **Develop capacity** at farm level to impart better knowledge in pest management in an integrated manner

☆ Intensive monitoring of crops at their vulnerable stages by effective means and linking it to **weather based** advisory system is essential

☆ Periodic pests and disease surveys to **update** the **knowledge** incidence, distribution, economic importance in different geographic regions with **farmer participation**

☆ Review of knowledge on **upcoming and new** species periodically to have substantial information ready on their management at all times

☆ Discuss and document the effective indigenous knowledge with added science on pest management that **augment natural enemies** should be of high priority

☆ Strategic research generated at the research stations need to be **evaluated and shared** periodically through farmer participatory approach.

☆ Establish and strengthen **farm clinics** for greater sustainability

☆ Registration, marketing and utilization of IPM inputs with reference to **bio-pesticides** need to be readdressed in order to encourage eco-friendly approaches for the benefit of environment and health

☆ Appropriate certification for IPM/**residue free products** should be put into practice with input and output market intelligence

☆ Address the post harvest losses to bridge the gap between the supply and demand.

Conclusions

Adequate support for plant protection research is essential to meet the challenge of understanding the population dynamics of key pests to produce healthy food from the available natural resources without any adverse effect on the environment. Technologies such as developing resistant varieties, enhancing natural enemies, improving the cultural control, judicious use of chemical pesticides under IPM will have significant role to play in future. Considering the ongoing changes in pest dynamics more research need to be concentrated on the basic biology of upcoming species to generate effective management strategies to be ready at every farm level. To achieve this periodic crop and pest surveys with the farmers perceptive need to be evaluated, updated and documented. This cannot be achieved through an individual research agenda of any organization over night; hence appropriate research partnerships including International organizations, national institutes, non-governmental agencies and farmers should be developed to provide better food to the growing populations.

References

Amit Sethi, Bons M S, Dilawari V K and Sethi A 2002. Response of whitefly *Bemisia tabaci* to selection by different insecticides and genetic analysis of attained resistance. *Resistant Pest Management Newsletter* **12**: 30-35.

Anonymous FAO: Ten years of IPM training in Asia. From farmer field school to community IPM. http://www.fao.org/docrep/005/ac834e/ac834e04.htm

Anonymous 2009. Biopesticides market to reach $ 1 billion in 2010. http://www.ien.com/article/biopesticides-market-to/8648.

Anonymous 2011. Indian Pesticides Industry – Vital for Ensuring Food Security. http://www.experiencefestival.com/wp/article/indian-pesticides-industry-vital-for-ensuring-food-security

Armes N J, Wightman J A, Jadhav D R and Ranga Rao G V 1997. Status of insecticide resistance in *Spodoptera litura* in Andhra Pradesh, India. *Pesticide Science* **50**: 240-248.

Berg H 2001. Pesticide use in rice and rice–fish farms in the Mekong Delta, Vietnam. *Crop Protection* **20**: 897-905.

Brethour C Weersink A 2003. Rolling the dice: on farm benefits of research in to reducing pesticide use. *Agricultural Systems* **76**: 575-587.

Budy P Resosudarmo 2001. Impact of the Integrated Pest Management Program on the Indonesian Economy. een.anu.edu.au/download files/een0102.pdf

Cuyno L C M, Norton G W and Rola A 2001. Economic analysis of environmental benefits of integrated pest management: a Philippine case study. *Agricultural Economics* **25**: 227-233.

Fernandez-Cornejo J 1998. Environmental and economic consequences of technology adoption: IPM in viticulture. *Agricultural Economics* **18**: 145-155.

Hammer G L, Jordan D R, Spiertz J H J, Struik P C and Laar H H van. Springer-Verlag GmbH, Heidelberg, Germany, Scale and complexity in plant systems research: gene-plant-crop relations, 2007, pp. 45-61.

Heinrichs E A and Mochida O 1984. From secondary to major pest status: the case of insecticide-induced rice brown plant hopper, *Nilaparvata lugens*, resurgence. *Protection Ecology* **7**: 201-218.

Jonathan A Foley, Navin Ramankutty, Kate A Brauman, Emily S Cassidy, James S Gerber, Matt Johnston, Nathaniel D Mueller, Christine O'Connell, Deepak

K Ray, Paul C West, Christian Balzer, Elena M Bennett, Stephen R Carpenter, Jason Hill1, Chad Monfreda, Stephen Polasky, Johan Rockstro, John Sheehan, Stefan Siebert, David and David P M Zaks 2011. Solutions for a cultivated planet: Nature/analysis, 20th October 2011, **478**: 337-342.

Joshi V, Ramana Sinha C S, Karuppaswamy M, Srivastava K K and Singh P B 1992. *Rural Energy Data Base*, TERI, New Delhi.

Kenis M and Cugala D 2006. Prospects for the biological control of the groundnut leaf miner, *Aproaerema modicella*, in Africa. Biocontrol News and Information. CAB Reviews: Perspectives in Agriculture, Veterinary Science, Nutrition and Natural Resources, 2006, 1, 031, 9 pp.

Kevin D Gallagher, Peter A C Ooi and Peter E Kenmore 2009. Impact of IPM Programs in Asian Agriculture. *In* Integrated Pest Management: Dissemination and Impact. *Springer Science + Business Media* **2**: 347-358.

Kranthi K R, Jadhav D R, Kranthi S, Wanjari R R, Ali S S and Russell D A 2002. Insecticide resistance in five major insect pests of cotton in India. *Crop Protection* **21**: 449-460.

Mac Hardy W E 2000. Current status of IPM in apple orchards. *Crop Protection* **19**: 801-806.

Mediene S, Valantin-Morison M, Sarthou J P,Tourdonnet S de, Gosme M, Bertrand M, Roger-Estrade, J, Aubertot J N, Rusch A, Motisi N, Pelosi C and Dore T 2011. Springer-Verlag, Paris, France, *Agronomy for Sustainable Development* **31**: 491-514.

Pagiola S 1995. Environmental and Natural Resource Degradation in Intensive Agriculture in Bangladesh, Paper No. 15 of the Environmental Economics Series, World Bank, June.

Pingali Prabhu L and Roberta Gerpacio 1997. Living with Reduced Insecticide Use for Tropical Rice. *Food Policy* **22**: 107-118.

Pinstrup Andersen P 2000. The future world food situation and the role of plant diseases. Website: http://www.scisoc.org/feature/Food Security/Top.html. protection measures: the IPM wheat model. *Plant Disease* **84**: 816-826.

Pratap S Birthal and Sharma O P 2004. Integrated Pest Management in Indian Agriculture: An Overview pp. 1-10 in Proceedings on Integrated Pest Management in Indian Agriculture (Ed. Pratap S Birthal and O P Sharma), Chandu Press D-97, Shakarpur Delhi - 110 092.

Ranga Rao G V and Rameswar Rao V 2010. Status of IPM in Indian Agriculture: A Need for Better Adoption. *Indian Journal of Plant Protection* **38**: 115-121.

Rola A C and Pingali P L1993. Pesticides, Rice Productivity and Health Impacts in the Philippines. In Paul Faeth (ed.) Agricultural Policy and Sustainability: Case Studies from India, Chile, the Philippines and the United States. Publisher World Resources Institute, Washington, D.C., U.S.A., pp. 47-62.

Schoenly K G, Cohen J E and Heong K L 1996. Quantifying the impact of insecticides on food web structure of rice arthropod populations in a Philippine farmer's irrigated field: A case study. 343–51. In Polis, G. and Winemiller, K. (eds.) *Food Webs, Integration of Patterns and Dynamics*. Chapman and Hall, London, UK.

Shivankar V J, Shyam Singh and Rao C N 2007. Secondary Pest Resurgence, Encyclopedia of pest management, Volume II, Chapter 155. Edited by David Pimentel CRC Press 2007, pp. 597-601.

Stefan Krull and Thies Basedow 2006. Studies on the biology of *Deanolis sublimbalis* Snellen (Lepidoptera: Pyralidae) and its natural enemies on mango in Papua New Guinea. http://www.dgaae.de/html/publi/mitt2006/273.pdf.

Tilman D 1999. Global environmental impacts of agricultural expansion: the need for sustainable and efficient practices. *Proceedings of National Academy of Sciences* **96:** 5995-6000.

Uaboi-Egbenni P O, Okolie P N, Okafor C N, Akinyemi O, Bisi-Johnson M A and O D Teniola 2010. Effect of soil types and mixtures on nodulation of some beans and groundnut varieties. *African Journal of Food Agriculture Nutrition and Development* **10:** 2272-2290.

United Nations 1996. World population prospects: The 1996 revisions. United Nations, New York.

Verreet J A, Klink H and Hoffmann G M 2000. Regional monitoring for disease prediction and optimization of plant protection measures. The IPM wheat model. *Plant Disease* **84:** 816-826.

Waterhouse D F 1998. Biological Control of Insect Pests: Southeast Asian Prospects. ACIAR Monograph No. **51:** 548 pp.

Way M J and van Emden H F 2000. Integrated pest management in practice-pathways towards successful application. *Crop Protection* **19:** 81-103.

Chapter 6

Application of Modern Tools of Biotechnology for Pest Management

Hari C Sharma

*International Crops Research Institute for the Semi-Arid Tropics (ICRISAT),
Patancheru – 502 324, Telangana State
E-mail: h.sharma@cgiar.org*

ABSTRACT

Recombinant DNA technology has significantly enhanced our ability for crop improvement, to meet the increased demand for food and fiber. Considerable progress has been made over the past two decades in manipulating genes from diverse sources to develop plants with resistance to insect pests, improve effectiveness of biocontrol agents, marker assisted selection for insect resistance, understand the nature of gene action and metabolic pathways, production of genetically modified sterile insects, use of molecular techniques in insect taxonomy, understand the mode of action of insecticides, and identify insecticides with newer mode of action. Despite the diverse and widespread beneficial applications of tools of biotechnology, there is a need to present these benefits to the public in a balanced manner. Testing and release of products generated through biotechnology-based processes should be continuously optimized based on experience. This will require a dynamic and streamlined regulatory structure, which is clearly supportive of the benefit of biotechnology, but highly sensitive to the well-being of humans and environment.

Keywords: Insect pests, Recombinant DNA technology, Marker-assisted selection.

Introduction

Nearly 30 to 50 per cent of the crop yields are lost due to the ravages of insect pests, and several insect species have the potential to cause 50 to 100 per cent loss during outbreaks. Insect pests cause an estimated annual loss of 13.6 per cent globally, and the extent of losses in India has been estimated to be 17.5 per cent (Dhaliwal *et al.*, 2010). The pest associated losses likely to increase as a result of changes in crop diversity and climate change. Reduction in pest associated losses is one of the potential areas for increasing food production, and it is in this context that we can exploit the tools of biotechnology to minimize the extent of losses due to insect pests. Molecular biology has provided several unique opportunities in crop improvement that include access to molecules novel to crop species, production of transgenic crops expressing insecticidal genes, ability to change the level of gene expression, and the capability to change the spatial and temporal pattern of gene expression (Sharma *et al.*, 2002). Development of effective insect-resistant varieties and biocontrol agents will lead to a reduced reliance on synthetic pesticides, and thereby reduce farmers' crop protection costs, while benefiting both the environment and public health. The promise of biotechnology for pest management can be realized by utilizing the information and products generated through research on genomics and genetic engineering to increase crop production.

Biotech Applications in Pest Management

Genetic Engineering

☆ Genetic engineering of crop plants for insect resistance.

☆ Genetic engineering of natural enemies.

☆ Genetic engineering of microbial pesticides.

☆ Genetic engineering of metabolic pathways.

☆ Inducible resistance and gene switches.

☆ Dominant repressible lethal genetic system.

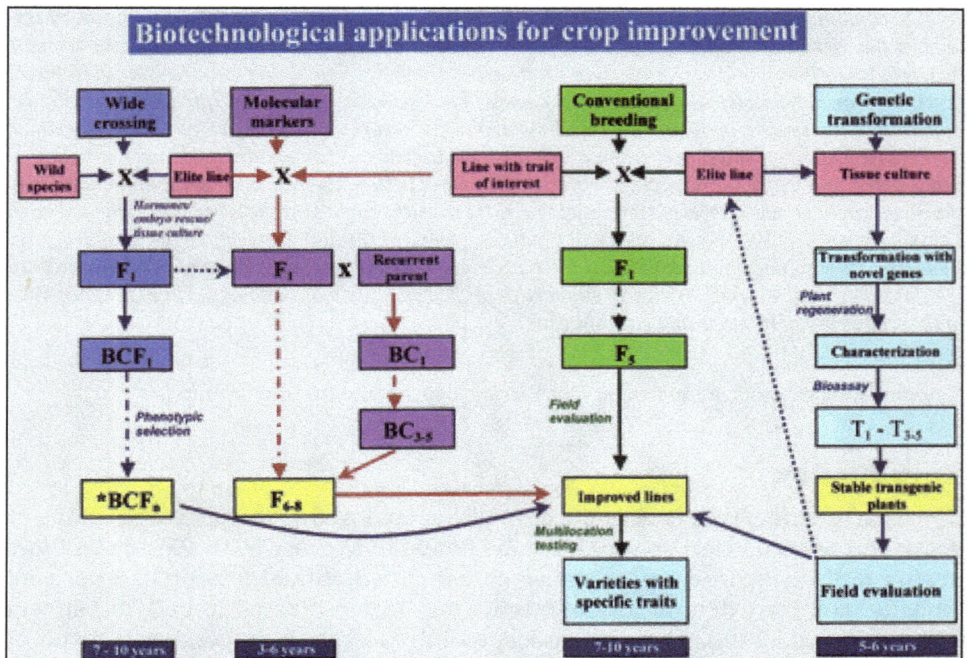

Biotechnological applications for crop improvement

Genomics and Molecular Markers

☆ Marker assisted selection for insect resistance.

☆ Diagnosis of insect pests and their natural enemies.

☆ Monitoring insect resistance to insecticides.

☆ Development of new pesticide molecules.

☆ Understand plant - insect - natural enemy interactions.

☆ Functional genomics and metabolisms of plants and insects.

Genetic Engineering

Crop Plants for Insect Resistance

Development and deployment of transgenic plants in an effective manner is an important pre-requisite for sustainable and economic use of biotechnology for crop improvement. As a result of advances in genetic transformation and gene expression over the past three decades, there has been rapid progress in using the tools of biotechnology for developing crops for resistance to insects (Sharma *et al.*, 2004). While most of the insect-resistant transgenic plants have been developed by using *Bt* δ-endotoxin genes, many studies are underway to use non-Bt genes, which interfere with development and the nutritional requirements of insects, including:

☆ **Cry toxins *Bt*:** Cry1Ab, Cry1Ac, Cry IIa, Cry9c, Cry IIB, Vip I, Vip II.

☆ **Plant metabolites:** Flavonoids, alkaloids, terpenoids.

☆ **Enzyme inhibitors:** SBTI, CpTi.

☆ **Enzymes:** Chitinase, lipoxygenase.

☆ **Plant lectins:** GNA, ACAL, WAA.

☆ **Toxins from predators:** Scorpion, spiders.

☆ **Insect hormones:** Neuropeptides and peptidic hormones.

Genes conferring resistance to insects have been inserted into maize, cotton, potato, tobacco, rice, broccoli, lettuce, walnuts, apples, alfalfa, and soybean (Sharma *et al.*, 2004). A number of transgenic crops have now been released for on-farm production or cultivation by the farmers (James, 2009). The first transgenic crop with resistance to insects was grown in 1994 (Benedict *et al.*, 1996). Since then, there has been a rapid increase in the area sown under transgenic crops and transgenic crops are now grown in over 20 countries in the world. Cry type toxins from *Bt* are effective against cotton bollworm, corn earworm, the European corn borer, and rice stem borers. Successful expression of *Bt* genes against the lepidopteran pests has also been achieved in tomato, potato, brinjal, groundnut, pigeonpea, and chickpea (Sharma, 2009). Development and deployment of transgenic plants is carried out under strict biosafety regulations in each country.

Deployment of insect-resistant transgenic plants should be based on the overall philosophy of integrated pest management, taking into account alternate mortality factors, reduction of selection pressure, and monitor insect populations for resistance development to design more effective management strategies. Trasngenic crops are compatible with other methods of pest control, and yield up to 50 per cent more than the non-transgenic cultivars even under insecticide protection (Sharma and Pampapathy, 2006). Insects such as *Heliothis virescens* (F.), *Helicoverpa zea* (Boddie), *Trichoplusia ni* (Hub.), and *Spodoptera exigua* (Hub.) are many times more sensitive to insecticide sprays when they have a prior exposure to *Bacillus thuringiensis* (Harris *et al.*, 1998). Transgenic crops can be used in conjunction with other methods of pest control without any detrimental or antagonistic effects. To increase the effectiveness and usefulness of transgenic plants, it is important to develop a strategy to minimize the rate of development of resistance in insect populations through:

☆ Control of secondary pests,

☆ Resistance management,

☆ Gene pyramiding and gene deployment,

☆ Regulation of gene expression,

☆ Planting refugia and destruction of carryover population,

☆ Control of alternate hosts and use of planting window, and

☆ Follow integrated pest management from the very beginning.

Metabolic Pathways

Genetic engineering can be used to change the metabolic pathways to increase the amounts of secondary metabolites, which play an important role in host-plant resistance to insect pests *e.g.*, medicarpin and sativan in alfalfa, maysin in maize, gossypol in cotton, stilbene in pigeonpea and chickpea, and deoxyanthocyanidin flavonoids (luteolinidin, apigenidin, etc.) in sorghum (Sharma *et al.*, 2002). The expression of phytoalexins in transgenic plants may be difficult due to complexities involved in their biosynthesis. Expression of a bacterial cytokinin biosynthesis gene (*PI-II-ipt*) in *Nicotiana plumbaginifolia* plants has been correlated with enhanced resistance to *M. sexta* and *M. persicae* (Smigocki *et al.*, 2000). Molecular mechanisms underlying the activation of defense genes implicated in phytoalexin biosynthesis are quite common in a large number of plant species. Biotechnology offers the promise to increase the production of secondary metabolites in plants to increase the levels of resistance to insect pests.

Inducible Resistance

Induced resistance results in changes in a plant that produce a negative effect on herbivores (Karban and Baldwin, 1997). Chemically induced expression systems or "gene switches" enable temporal, spatial, and quantitative control of genes introduced into plants or those that are already present in the plants to impart resistance to insects. A number of inducible genes have been identified in plants based on endogenous chemical signals such as phytohormones, responses to insect and pathogen attack, or wounding. Effectiveness of the chemical injury inducer, Actigard™ in providing resistance to various insect pests and pathogens in the tomato has been demonstrated by Inbar *et al.* (1998). Proteinase inhibitors and oxidative enzymes such as polyphenol oxidase, peroxidase, and lipoxygenase persist for at least 21 days after induction in damaged tomato leaflets (Stout *et al.*,

1996). Exogenous application of jasmonic and salicylic acids induces resistance to several insect pests (War *et al.*, 2012).

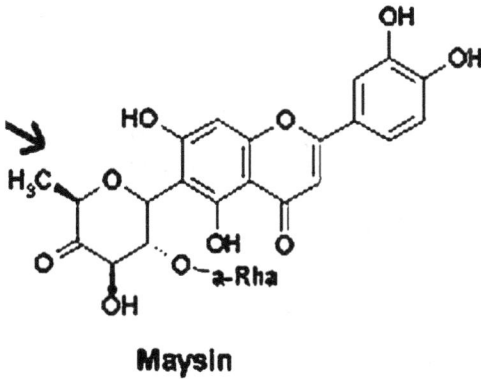

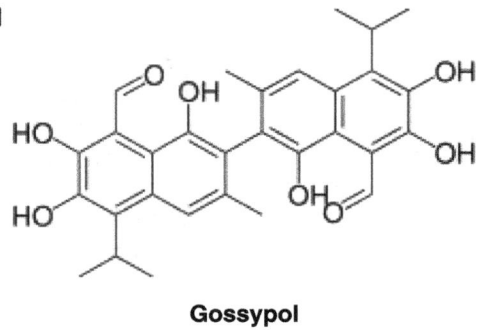

Maysln   **Gossypol**

Natural Enemies

Some of the major problems in using natural enemies in pest control are the difficulties involved in mass rearing and their inability to withstand adverse conditions in the field. Genetic improvement can be useful when the natural enemy is known to be a potentially effective biological control agent, except for one limiting factor. Some of the desirable characteristics for transgenic insects include resistance to pathogens, adaptation to different environmental conditions, high fecundity, and improved host-seeking ability (Atkinson and O'Brochta, 1999). Biotechnological interventions can also be used to broaden the host range of natural enemies or enable their production on artificial diet or non-host insect species that are easy to multiply under laboratory conditions. In addition, there is a tremendous scope for developing natural enemies with genes for resistance to pesticides (Hoy, 2000). This is of particular concern when the same vector transmits several disease causing pathogens, as it might be difficult to develop transgenic individuals incapable of transmitting different pathogens (Sharma, 2009).

Microbial Pesticides

Genetic engineering can also be used to improve the efficacy of entomopathogenic microorganisms. Efforts to improve *Bt* have largely been focused on increasing its host range and stability. Work on *baculoviruses* is largely focused on incorporation of genes that produce the proteins, which kill the insects at a faster rate (Bonning and Hammock, 1996), and removal of polyhedrin gene, which produces the protective viral-coat protein, and its persistence in the field (Cory, 1991). Neurotoxins produced by spiders and scorpions have also been expressed in transgenic organisms (Barton and Miller, 1991). Incorporation of benomyl resistance into *Metarhizium anisopliae* and other entomopathogenic fungi could make them more useful for use in integrated pest management (Goettel *et al.*, 1989). The role of neurotoxins from insects and spiders need to be studied in greater detail before they are deployed in other organisms because of their possible toxicity to mammals.

Dominant Repressible Lethal Genetic System

The sterile insect technique has been employed to control several insect pests. However, this system depends on large-scale production of the target insect, and use of irradiation or chemical sterilization. Release of insects carrying a dominant lethal (RIDL) gene has been proposed as an alternative to the conventional techniques used for insect sterilization (Alphey and Andreasen, 2002). This is based on the use of a dominant, repressible, female-specific gene for insect control. A sex-specific promotor or enhancer gene is used to drive the expression of a repressible transcription factor, which in turn controls the production of a toxic gene product. A non-sex specific expression of the repressible transcription factor can also be used to regulate a selectively lethal gene product. Insects produced through genetic transformation using this approach do not require sterilization through irradiation, and could be released in the eco-system to mate with the wild population to produce sterile insects, which will be self-perpetuating.

Genomics and Molecular Markers

Marker-Assisted Selection for Insect Resistance

Recombinant DNA technologies, besides generating information on gene sequences and function, allow the identification of specific chromosomal regions carrying genes contributing to traits of economic interest. Once genomic regions contributing to the trait of interest have been assigned and the alleles at each locus designated, they can be transferred into locally adapted high-yielding cultivars by making requisite crosses. The offspring with a desired combination of alleles can then be selected for further evaluation using marker assisted selection (MAS). It is important to use large mapping populations, which are precisely and accurately characterized across seasons and locations. MAS can be used to accelerate the pace and accuracy of transferring insect resistance genes into improved cultivars. Several markers have been used to identify QTLs for insect resistance in different crops (Smith, 2005; Sharma, 2009). In contrast to the markers linked to resistance genes inherited as simple dominant traits, the improvement of polygenic traits (QTLs) through MAS is difficult due to involvement of a number of genes, and their interactions (epistatic effects). Several studies on QTLs linked to stem borer resistance in maize underscore the problems involved in using QTLs in MAS. The relative efficiency of phenotypic and MAS has been found to be similar). However, phenotypic selection was more favorable due to lower costs. Maximum progress has been made in breeding for insect resistance in common bean by using a combination of phenotypic performance and QTL-based index, followed by QTL based index, and conventional selection (Tar'an *et al.*, 2003).

Understanding Gene Sequence and Function

Genes can be discovered using a variety of approaches (Shoemaker *et al.*, 2001), but a routine large-scale approach can commonly be followed by generating and sequencing a library of expressed genes. A large number of ESTs are now available in the public databases for several crops such as *Zea mays*, *Arabidopsis thaliana*, *Oryza sativa*, *Sorghum bicolor*, and *Glycine max*. A comparison of the EST databases from different plants can reveal the diversity in coding sequences between closely and distantly related species, while mapping of ESTs may elucidate the synteny between those species. For understanding gene functions of a whole organism, functional genomics technology is now focused on high throughput methods using insertion mutant isolation, gene chips or microarrays, and proteomics. These and other high throughput techniques offer powerful new uses for the genes discovered through sequencing (Hunt and Livesey, 2000).

DNA Barcoding of Insect Pests and their Natural Enemies

For developing appropriate strategies for managing insect pests, it is important to have a correct identification of the pest species. Correct taxonomic identification is also important for import and export of plant material/food grains to implement quarantine procedures. Identification of insect pests has primarily relied on morphological characters of adult life stages. However, intercepted specimens often are not in the adult stage and may be damaged, which seriously handicaps correct identification. The molecular tools now enable precise and rapid identification of insect pests, irrespective of the developmental stage and condition of the samples. The modern tools of biotechnology can be used for detection and identification of insect pests, insect biotypes, and understand genetic diversity, population structure, tri-trophic interactions, and insect plant relationships (Caterino *et al.*, 2000; Heckel, 2003). Molecular markers can also be used to gain a basic understanding of their interaction with environment, and develop sound strategies for pest management.

Development of New Insecticide Molecules

Crop protection is still dominated by chemical control, and this approach will continue to be important in crop protection in future. Traditionally, the discovery of new agrochemicals has used *in vivo* screens to identify new compounds. Functional genomics offers the opportunity to acquire in-depth knowledge of the genetic make-up and gene function of insect pests that may lead to the discovery of new processes that could be the targets for novel chemistry (Hess *et al.*, 2001). Combining genomics with high throughput biochemical screening can be used to identify a range of new chemicals for pest control. Genomic technologies are now allowing investigation of some previously intractable mechanisms involved in insect resistance to insecticides. New molecular techniques permit fundamental insights into the nature of mutations and genetic processes such as gene amplification, altered gene transcription, and amino acid substitution to underpin insecticide resistance mechanisms. This in turn will lead to high-resolution diagnostics for resistance alleles in homozygous and heterozygous form, especially for insect pests with multiple resistance mechanisms, or for resistance mechanisms not amenable to biochemical assays.

Large-scale adoption of insect-resistant transgenic crops has resulted in a significant reduction in insecticide use and increased both production and productivity (Qaim and Ziberman, 2003; James, 2009). The potential of insect-resistant transgenic crops can be enhanced through gene pyramiding by using a combination of exotic genes and insect-resistant cultivars derived through conventional breeding, and by combining resistance to insect pests and diseases of importance in a crop/region. There is a considerable debate about the environmental risks such as development of resistance, harmful effects on beneficial insects, and gene flow to the closely related wild relatives of the crops (Sharma and Ortz 2000; O'Callaghan *et al.,* 2005; Sharma, 2009; Sharma *et al.,* 2012). The evidence on these issues is still inconclusive and warrants careful monitoring before transgenic crops are deployed on a large scale. There is a need for a balanced presentation of the benefits of biotechnology to the general public for increasing crop production and improving food security. The biggest risk of modern biotechnology for developing countries is that technological developments may bypass the poor farmers because of a lack of enlightened adoption. There is a need to develop scientifically sound strategies for deploying genetically engineered insect-resistant crops for sustainable crop production. Equally important is the need to assess the bio-safety of genetically modified crops and the conventional technologies deployed for pest management.

References

Alphey L and Andreasen M 2002. Dominant lethality and insect population control. *Molecular and Biochemical Parasitology* **121**: 173-178.

Atkinson P W and O'Brochta D A 1999. Genetic transformation of non-drosophilid insects by transposable elements. *Annals of the Entomological Society of America* **92**: 930-936.

Barton K A and Miller M J 1991. Insecticidal toxins in plants. *EPA 0431829 A1 910612.*

Benedict J H, Sachs E S, Altman D W, Deaton D R, Kohel R J, Ring D R and Berberich B A 1996. Field performance of cotton expressing CryIA insecticidal crystal protein for resistance to *Heliothis virescens* and *Helicoverpa zea* (Lepidoptera: Noctuidae). *Journal of Economic Entomology* **89**: 230-238.

Bonning B C and Hammock B D 1996. Development of recombinant baculovirus for insect control. *Annual Review of Entomology* **41**: 191-210.

Caterino M S, Cho S and Sperling, FA 2000. The current state of insect molecular systematics: a thriving Tower of Babel. *Annual Review of Entomology* **45**: 1-54.

Cory J S 1991. Releases of genetically modified viruses. *Medical Virology* **1**: 79-88.

Dhaliwal G S, Jindal V and Dhawan A K 2010. Insect pest problems and crop losses: Changing trends. *Indian Journal of Ecology* **37**: 1-7.

Goettel M S, St Leger R J, Bhairi S, Jung M K, Oakley, B R and Staples R C 1989. Transformation of the entomopathogenic fungus, *Metarhizium anisopliae,* using the *Aspergillus nidulans* ben A3 gene. *Current Genetics* **17**: 129-132.

Harris J G, Hershey C N, Watkins M J and Dugger P 1998. The usage of Karate (lambda-cyhalothrin) oversprays in combination with refugia, as a viable and sustainable resistance management strategy for B.T. cotton. In: *Proceedings, Beltwide Cotton Conference,* 5-9 January 1998, San Diego, California, USA. Volume 2. Memphis, USA: National Cotton Council. pp. 1217-1220.

Heckel DG 2003. Genomics in pure and applied entomology. *Annual Review of Entomology* **48**: 235-260.

Hess F D, Anderson R J and Reagan J D 2001. High throughput synthesis and screening: the partner of genomics for discovery of new chemicals for agriculture. *Weed Science* **49**: 249-256.

Hoy M A 2000. Transgenic arthropods for pest management programs: risks and realities. *Experimental and Applied Acarology* **24**: 463-495.

Hunt S P and Livesey F J 2000. *Functional Genomics: A Practical Approach.* Oxford University Press, Oxford. 253 pp.

Inbar M, Doodstar H, Sonoda R M, Leibee G L and Mayer R T 1998. Elicitors of plant defensive systems reduce insect densities and disease incidence. *Journal of Chemical Ecology* **24**: 135-149.

James C 2009. Global status of commercialized biotech/GM crops: 2008. *ISAAA Brief No. 39.* International Service for the Acquisition of Agri-Biotech Applications, ISAAA: Ithaca, NY, USA.

Karban R and Baldwin I T 1997. *Induced Responses to Herbivory.* Chicago, Illinois, USA: The University of Chicago Press.

O'Callaghan M, Glare T R, Burgess E P J and Malone L A 2005. Effects of plants genetically modified for insect resistance on nontarget organisms. *Annual Review of Entomology* **50**: 271-292.

Qaim M and Zilberman D 2003. Yield effects of genetically modified crops in developing countries. *Science* **299**: 900-902.

Sharma H C 2009. *Biotechnological Approaches for Pest Management and Ecological Sustainability.* Boca Raton, Florida, USA: CRC Press. 526 pp.

Sharma H C and Ortiz R 2000. Transgenics, pest management, and the environment. *Current Science* **79**: 421-437.

Sharma H C and Pampapathy G 2006. Influence of transgenic cotton on the relative abundance and damage by target and non-target insect pests under different protection regimes in India. *Crop Protection* **25**: 800-813.

Sharma H C, Crouch J H, Sharma K K, Seetharama N and Hash C T 2002. Applications of biotechnology for crop improvement: prospects and constraints. *Plant Science* **163**: 381-395.

Sharma H C, Dhillon M K and Sahrawat K L 2012. *Environmental Safety of Biotech and Conventional IPM Technologies.* Studium Press LLC, P.O. Box 722200, Houston, TX 77072, USA. 426 pp.

Sharma H C, Sharma K K and Crouch J H 2004. Genetic engineering of crops for insect control: Effectiveness and strategies for gene deployment. *CRC Critical Reviews in Plant Sciences* **23**: 47-72.

Shoemaker D D, Schadt E E, Armour Y D, He P, Garrett-Engel P D and McDonagh P M 2001. Experimental annotation of the human genome using microarray technology. *Nature* **409**: 922-927.

Smigocki A, Heu S and Buta G 2000. Analysis of insecticidal activity in transgenic plants carrying the ipt plant growth hormone gene. EUCARPIA TOMATO 2000.

XIV meeting of the EUCARPIA Tomato Working Group, Warsaw, Poland, 20-24 August 2000. *Acta Physiologia Plantarum* **22**: 295-299.

Smith C M 2005. *Plant Resistance to Arthropods – Molecular and Conventional Approaches.* Dordrecht, The Netherlands: Springer Verlag. 423 pp.

Stout M J, Workman J and Duffey S S 1996. Differential induction of tomato foliar proteins by arthropod herbivores. *Journal of Chemical Ecology* **20**: 2575-2594.

Tar'an B, Thomas E, Michaels T E and Pauls K P 2003. Marker assisted selection for complex trait in common bean (*Phaseolus vulgaris* L.) using QTL-based index. *Euphytica* **130**: 423-433.

War A R, Paulraj M G,Tariq A, Buhroo A A, Ignacimuthu S and Sharma H C 2012. Mechanisms of plant defense against insect herbivores. *Plant Signaling and Behavior* **7**: 1306-1320.

Chapter 7

Nanotechnology Applications in Plant Health Management

K S Subramanian, K Gunasekaran, M Kannan and M Praghadeesh

Department of Nano Science and Technology,
Tamil Nadu Agricultural University,
Coimbatore – 641 003, Tamil Nadu
E-mail: kssubra2001@rediffmail.com

ABSTRACT

Nanotechnology is a fascinating field of science, which manipulates atom by atom and thus processes and products evolved are the most précised ones that are impossible to achieve by the conventional systems. In plant health management, nanotechnological principles and concepts can be exploited for early detection of pests, diseases and nutrient deficiencies, smart delivery of agricultural inputs besides surveillance. Protection of soil and plant health requires rapid, sensitive detection of pollutants and pathogens with molecular precision. The nanotechnology applications in plant health management include (i) **Biosensors** for early detection of major pests (eg. eriophid mite, mealy bugs, cotton weevil etc.), diseases (e.g. red rot in sugarcane, downy mildew in grapes) and abiotic stresses (drought, salinity, Zn deficiency), (ii) **Nano-pheromones** with a sustained release of semio-chemicals to achieve effective pest control, (iii) **Nano-encapsulation** of pesticides for regulated release for specific needs, (iv) **Metal-oxide nano-particles** for plant protection besides (v) **Smart delivery system** for agri-inputs. Our literature review clearly suggests that there is an abundance of scope to exploit nanotechnology in plant health management through smart delivery systems which facilitates enhanced use efficiency of inputs with environmental protection. It is time that agricultural scientists should undertake research in the field of nano-based smart delivery systems to achieve the targeted delivery of inputs that enhances the crop productivity with minimal use agri-inputs.

Keywords: Nano-biosensor, Pheromones, Nano-encapsulation, Nano-pesticide, Smart delivery system.

Introduction

Indian agriculture has undergone a series of transformations that led to the paradigm shift from traditional farming to precision agriculture. In the past decade, the agricultural production system has faced a wide array of challenges that include burgeoning population, shrinking farm land, restricted water availability, imbalanced crop nutrition, multi-nutrient deficiencies in crops, yield stagnation and decline in organic matter. To address all the challenges ahead, we should think of an alternate technology such as "nanotechnology" to precisely detect and deliver the correct quantity of agri-inputs required by crops that promote productivity with environmental safety.

Nanotechnology is a fascinating field of science, which manipulates atom by atom and thus processes and products evolved are the most precised ones that are impossible to achieve by the conventional systems. Nano-particles measure a dimension of 10^{-9} m *i.e.* billionth of a metre. For instance, a virus particle may be sliced into 100 nano-particles or 80,000 nano-particles can be arranged across a human hair. Simply, consider that entire Indian population of 1.20 billion people assembled in 1m length, each Indian is a dimension of a nano-particle. Since, the nano-particles are extremely small, their surface mass ratio is huge that facilitates manipulation at the atomic scale (Figure 7.1). Nano-particles exhibit different physical strength, chemical reactivity, electrical conductance and magnetic properties (Nykypanchuk *et al.*, 2008).

Nanoscience infuses intelligence to the truck load of chemical constituents that are to be delivered at appropriate locations and cleave from the site after the task is complete. Such process will likely to reduce the cost besides ensuring environmental safety. Nanoscale devices with its unique properties make the agricultural system more smart and effective; such devices are capable of responding to different situations by themselves, thus taking appropriate remedial action with the need of external directions from humans. In short, these devices act as a detectors and if need arises serve as a solution/remedy for the particular issue. These smart delivery systems of chemicals in controlled and targeted manner can be considered synonymous to the proposed nano-drug delivery system in human (Patolsky *et al.*, 2006). Despite nanotechnology being exploited in the fields of electronics, energy and health sectors, agricultural science is just beginning to scratch the surface. Nanotechnology is being visualized as a rapidly evolving field that has potential to revolutionize agriculture and food systems and improve the conditions of the poor. The Indian Council of Agricultural Research (ICAR) has introduced "Nano Mission in Agriculture" as one of the Platforms to be implemented during the 12th Five year plan period (2012-2017).

Biosensors and Diagnostic Kits for Early Detection

Pests, diseases and nutrient deficiencies constitute major loss to the tune of 40-65 per cent of any agricultural or horticultural crops. Early detection is utmost essential to protect the crops from infection and prevent yield and quality losses. Conventionally, pesticides are sprayed only after symptoms are obvious based on visual diagnosis. When spraying is performed it may be too late to protect the

8 Cubes Side L
Each has Surface area 6L^2
Total Surface Area 48 L^2

1 Cube
Length of sides 2L
Surface area 24 L^2

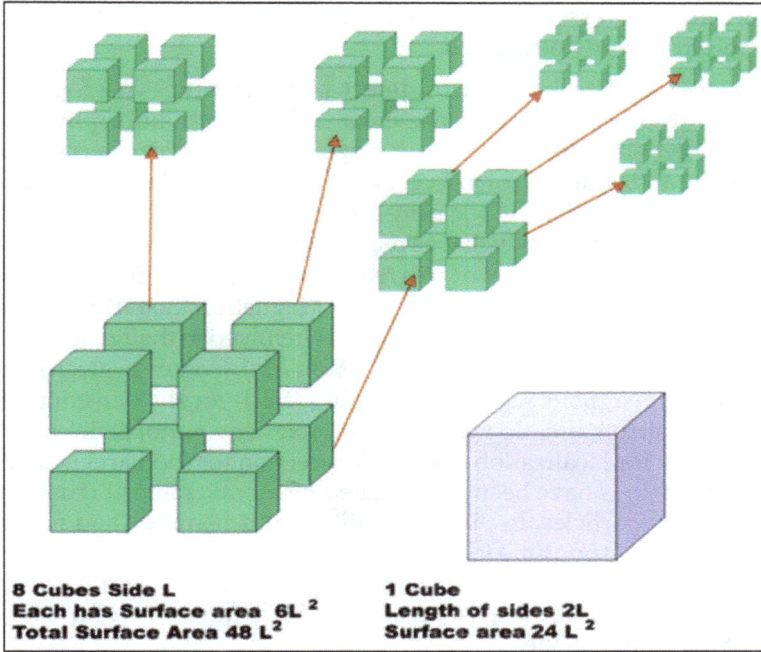

Figure 7.1: Unique Property of Nano-Particles – Extensive Surface Area.

crops. Biosensors can be developed in order to accurately measure the moisture, nutrients, pathogenicity and pest incidence so as to take up timely corrective measures. Development of biosensors for major pests (*e.g.* Eriophid mite, mealy bugs, cotton weevil etc.), diseases (*e.g.* red rot in sugarcane, downy mildew in grapes) and abiotic stresses (drought, salinity, Zn deficiency) will be an ideal tool to protect the crop from devastation.

Biosensors can also be used to detect pest incidence in crops. Magnetic nanoparticles are known to be omnipresent and their distribution pattern could be used for the detection of pest incidence in crops. The magnetic material is present in the head, thorax and abdomen of insects like *Solenopsis substitute* (Fabricius), an ant. These magnetic nanoparticles in social insects act as geomagnetic sensors (Esquivel *et al.*, 2007). The observation which was made through electron microscopy, clearly demonstrated that several species of ants recognize magnetic signals with the help of magnetite nanoparticles (Abracado *et al.*, 2005). Magnetic nanoparticles in the honey bee, *Apis mellifera* abdomens are well accepted as involved in their magnetoreception mechanism (El-Jaick *et al.*, 2001). Fire ant (*S. invicta*) workers, queens and alates were analyzed by magnetic resonance imaging (MRI) for the detection of natural magnetism. All ferromagnetic materials are magnetic nanoparticles, which are solely responsible for localization of specific direction for food and host of insects. Recently, cicada wings have been investigated by atomic force microscopy (AFM) for observing nanoparticles. In plants, volatile phytochemicals and nanoparticles of insects are solely responsible for plant-insect interaction (Gorb and Gorb, 2009). Nanoscience strongly suggests that nanoparticles can be exploited as a hue to assess the occurrence of insect pests which facilitate quick reaction time to sense and prevent the pest damage.

In nanomechanical biosensors, receptor molecules are immobilized on the surface of a microcantilever such as those used in atomic force microscopy but without scanning probes. The most common method is measuring the deflection of a cantilever, in which only a single side is coated with receptor molecules. Molecular recognition on the sensitized cantilever side, gives a change of the surface stress due to the electrostatic, van der Waals, configurational and stearic interactions between the adsorbed molecules. This technique can be exploited to detect pests and diseases (Franca *et al.*, 2011).

Nanotechnological approaches are widely used for early detection of diseases particularly cancer in humans. Similar diagnostic approaches and devices can be exploited in agricultural production systems. In the past two decades, viral diseases in several crops can be detected using ELISA (Enzyme linked immunosorbent assay) tests. This method is based on antigen- antibody reaction that is very specific and accurately detects the diseases. Despite this technique is highly useful, it takes a couple of weeks to get the ELISA tests done in nearby plant pathological laboratories by that time extensive damage could have already been done and it becomes too late to undertake control measures. The ICAR has taken assiduous efforts to undertake research on dip-stick method wherein proteins or nucleic acids serve as reference molecules. The plant extract is allowed to react with the stick and the detection is done within couple of minutes in the field itself. This technique has been proved effective in detecting viral diseases in potato. The precision and validity of the method can be further improved using nano-particles.

Recently, Dr. Florian Schröper of the Fraunhofer Institute for Molecular Biology and Applied Ecology IME in Aachen, Germany, has developed a quick test kit to provide the farmer a low-cost pathogen detection right there in the field. At the heart of the test is a magnetic reader devised by scientists at the Peter Grünberg Institute of the Forschungszentrum Jülich. The device has several excitation and detection coils arrayed in pairs. The excitation coils generate a high- and low-frequency magnetic field, while the detection coils measure the resulting mixed field. If magnetic particles penetrate the field, the measuring signal is modified. The result is shown on a display, expressed in millivolts. This permits conclusions about the concentration of magnetic particles in the field. Researchers are making use of this mechanism to track down pathogens. "What they detect is not the virus itself but the magnetic particles that bond with the virus particles. These are first equipped with antibodies so that these can specifically target and dock onto the pathogens. This way, essentially there is a virus particle "stuck" to each magnetic particle. To ensure that these are in proportion to one another, researchers use a method that functions similar to the ELISA principle. They introduce plant extract into a tiny filtration tube filled with a polymer matrix to which specific antibodies were bound. When the plant solution passes through the tube, the virus particles are trapped in the matrix. Following a purification step, the experts add the magnetic particles modified with antibodies. These, in turn, dock onto the antigens in the matrix. A subsequent purification step removes all of the unbound particles. The tube is then placed in an appliance in the magnet reader to measure the concentration of magnetic particles. The researchers have already achieved promising results in initial tests involving the grapevine virus: the measured values reached a level of sensitivity ten times as that of the ELISA method. Currently, Schröper and his team are working to expand their tests to other pathogens such as the mold spore *Aspergillus flavus*. The development of diagnostic kits for detection of diseases, pests and nutrient deficiencies are quite appropriate in the context of Indian Agriculture to enable quick reaction time to take corrective measures.

Nano-Pheromones

One of the immediate possibilities of using nanoscience in pest control is the pheromones. These compounds are highly volatile and their release pattern can be regulated through nano-formulations. In this technology, pheromone lure compound is kept in a trap to attract female insects and kill them for the effective control of pests. Research has been done for decades for the isolation, identification and synthesis of insect pheromones which is an effective tool for management of pest as one the IPM strategy but still the technology needs refinement for use in the applications of mass trapping and mating disruption with that ensures formulations possessing, less photo-oxidation, least thermo degradation and prolonged shelf life. This study focuses on the synthesis of nano-pherosensor (nano-sensor used to sense pheromone) to detect pheromone molecules and provide concentration present in field. After recognition of pheromone molecules this should act as trigger for device which should release optimum level of pheromones.

In India, pheromone lures are used for trapping fruit flies, stem borers in sugarcane, boll worms *Heliothis armigera* and *Spodoptera litura*, coconut red palm weevil, *Rhychophorus ferrugineus* and rhinoceros beetle, *Oryctes rhinoceros*. In certain cases like the banana stem weevil, *Odoiporous longicollis* and sweet potato weevil,

Cylas formicarius. They are yet to be used on wide scale in field level due to non availability of a robust delivery system. For pests like cashew stem borer, citrus leaf miner and oilpalm psychid, *Metesia plana*, the compounds that cause physiological and behavioural response needs to be identified. Recently, Bhagat *et al.* (2013) has developed a pheromone-based nano-gel in order to regulate the release of methyl euginol against fruit fly (*Bactrocera dorsalis*) in guava. The nano-gel has been developed using supramolecular principles and nanotechnological approaches that are chemically, mechanically and thermally stable and offered extended release of euginol to the tune of 30 weeks under field conditions while such release ceased to exist within three weeks. In Tamil Nadu Agricultural University, a mixture of pheromone molecules (Z) – 11 hexa decenal and (Z) – 9 – hexadecenal against rice yellow stem border (*Scirpophaga incertulas*) has been successfully regulated through a zeolite (369 nm) substrate (Prakash, 2013). This fortification process assisted in regulated release from 9 to 63rd day for the moth catches while the conventional rubber septa attracted the moth from 1 to 24 days. In another approach, sex pheromone of European grapevine moth (*Lobesia botrana*) (E,Z) – 7,9-dodecadien-l-yl acetate (DA) was immobilized in an electrospun fibe matrix made of oligo-lactide (OLA) (Bansal *et al.*, 2012). Such process facilitates regulated release of the pheromone molecules. Mihou *et al.* (2006) have shown that the polyurea capsules can be used as a carrier to slow down the release of (Z) – 11 hexadecenyl acetate which serves as the pheromone system for a wide array of Noctuidae species. The robustness of the chemo-ecological approach for pest management can be achieved by studying the interactions between organisms belonging to different trophic levels at a broader level. The results of such studies will pave way for environmentally sound practices of pest monitoring and control, as the plant derived compounds and species specific pheromones may interact synergistically in attracting targeted insect pest. The success of the pheromone/kairomone technology depends of development of an effective sustainable delivery system. Development of a nano-matrix for release of pheromone/kairomone to be taken up under network mode will help to develop a clean technology for management of key pests of crops.

Nano Encapsulation of Pesticides

Persistence of insecticides in the initial stage of crop growth helps in bringing down the pest population below economic threshold level and to have an effective control for a longer period. Hence, the use of residues in the applied surface remains one of the most cost-effective and versatile means of controlling insect pests. In order to protect the active ingredient from the environmental conditions and to promote persistence, a nanotechnology approach "microencapsulation" can be used to improve the insecticidal value. Microencapsulation comprises nano-sized particles of the active ingredients being sealed by a thin-walled sac or shell (protective coating). Microcapsules generally measure from 5 to 500 microns in size. In Tamil Nadu Agricultural University, neem-based microemulsion (198 nm) has been developed and found effective in controlling sucking pests such as thrips, aphids and mites (Gunasekaran, 2011). Recently, several research papers have been published on the encapsulation of insecticides. Nanoencapsulation with nanoparticles in form of pesticides allows for proper absorption of the chemical into the plants unlike the case of larger particles (Scrinis and Lyons, 2007). Nano-encapsuation of insecticides, fungicides or nematicides will help in producing nano-formulations which offer effective control of pests while preventing residues in soil.

Similar to insecticides or fungicides, herbicides have been encapsulated to suit the agro-climatic systems particularly rainfed. Under rainfed conditions, there is no guarantee for moisture availability and thus herbicides are to be designed and fabricated to release the active ingredient only when the soil receives a short spell of rainfall. Nano-capsulated herbicides are known to control the notorious parasitic weeds while reducing the phytotoxicity of herbicides on crops that explains the benefits of smart delivery system in agriculture. Properly functionalized nano-capsules provide better penetration through cuticle and allow slow and controlled release of active ingredients on reaching the target weed. Nano encapsulation of chemicals with biodegradable materials makes safer and easy to handle by the growers. Efforts are underway to kill the notorious weeds like *Cyprus rotundus* through smart delivery system. This weed produces tubers rich in starch that has to be exhausted through a suitable smart delivery system (Perez-de-luque and Diego, 2009; Kanimozhi and Chinnamuthu, 2012). Nano-encapsulated agrochemicals should be designed in such a way that they possess all necessary properties (effective concentration, stability and solubility) time controlled release in response to certain stimuli, enhanced targeted activity and less eco-toxicity with safe and easy mode of delivery thus avoiding repeated applications.

Nano-particles for Plant Protection

Nano-particles are being used as effective strategy to protect the crops from damaged by pests and diseases. Surface modified hydrophobic nanosilica has been successfully used to control a range of agricultural pests (Barik *et al.*, 2008). This functionalized lipophilic nanosilica gets absorbed into the circular lipids of insects by physiosorption and damages the protective wax layer and induces death by desiccation. The use of such nano-biopesticide is more acceptable since they are safe for plants and cause less environmental pollution in comparison to conventional chemical pesticides (Rahman *et al.*, 2009).

The successful use of silver nanoparticles (Ag NPs) in diverse medical streams as antifungal and antibacterial agents has led to their applications in controlling phytopathogens. Ag NPs with broad spectrum of antimicrobial activity reduce various plant diseases caused by spore producing fungal pathogens. The effectiveness of Ag NPs can be improved by applying them well before the penetration and colonization of fungal spores within the plant tissues (Singh *et al.*, 2008). The small size of the active ingredient effectively controls fungal diseases like powdery mildew. However, it was also observed that a very high concentration of nano-silica-silver produced some chemical injuries on the cucumber. The use of Ag NPs as an alternative to pesticides for the control of sclerotium forming phytopathogenic fungi was also investigated. Exposure of fungal hyphae to Ag NPs caused severe damage by the separation of layers of hyphal wall and collapse of hyphae. The efficacy of Ag NPs in extending the vase life of gerbera flowers was also studied and the results show inhibited microbial growth and reduced vascular blockage which increased the water uptake and maintained the turgidity of gerbera flowers (Solgi *et al.*, 2009). Apparently, the use of biocide containing polymeric nanoparticles for introducing organic wood preservatives and fungicides into wood products thereby reducing the wood decay was also studied (Liu *et al.*, 2001). Among the nano-particles, Ag NPs are widely used accounting for more than 30 per cent of the nano-based commercial products in the world. The use of nano-particles in plant protection and production is summarized in Table 7.1.

Table 7.1: Use of Nano-Particles in Agro-ecosystems

Application	Nanoparticles	Reference
Pesticide delivery		
Avermectin	Porous hollow silica (15 nm)	Li *et al.*, 2007
Ethiprole or phenylpyrazole	Poly-caprolacetone (135 nm)	Boehm *et al.*, 2003
Gamma cyhalothrin	Solid lipid (300 nm)	Frederiksen *et al.*, 2003
Tebucanazole/chlorothalonil	Polyvinylpyridine and polyvinyl-pyridine-co-styrene (100 nm)	Liu *et al.*, 2001
Biopesticide		
Plant origin: nanosilica for insect control *Artemisia arborescens*	Nanosilica (3-5 nm)	Barik *et al.*, 2008
Essential oil encapsulation	Solid lipid (200-294 nm)	Lai *et al.*, 2006
Microorganisms: *Lagenidium giganteum* cells in emulsion	Silica (7-14 nm)	Vandergheynst *et al.*, 2007
Fertilizer delivery		
NPK controlled delivery	Nano-coating of sulfur (100 nm layer)	Wilson *et al.*, 2008
	Chitosan (78 nm)	Corradini *et al.*, 2010
Genetic material delivery (DNA)	Gold (10-15 nm)	Torney *et al.*, 2007
	Gold (5-25 nm)	Vijayakumar *et al.*, 2010
	Starch (50 – 100 nm)	Liu *et al.*, 2008
Double stranded RNA	Chitosan (100 – 200 nm)	Zhang *et al.*, 2010
Pesticide Sensor		
Carbofuran/triazophos	Gold (40 nm)	Guo *et al.*, 2009
DDT	Gold (30 nm)	Lisa *et al.*, 2009
Dimethoate	Iron oxide (30 nm)	Gan *et al.*, 2010
Organophosphate	Zirconium oxide (50 nm)	Wang *et al.*, 2009
Pyrethroid	Iron oxide (22 nm)	Kaushik *et al.*, 2009
Pesticide degradation		
Lindane	Iron sulfide (200 nm)	Paknikar *et al.*, 2005
Imidacloprid	Titanium oxide (30 nm)	Gaun *et al.*, 2008

Nano Fertilizers for Balanced Crop Nutrition

Nanotechnology is currently being exploited for the regulated release of nutrients from the fertilizers. Nanofertilizers are designed to deliver nutrients slowly and steadily matching with the crop requirement. This can be achieved by preventing nutrients from interacting with soil, water and microorganisms, and releasing nutrients only when they can be directly internalized by the plant (De Rosa *et al.*, 2010). Liu *et al.* (2006) have shown that coating and cementing of nano and sub- nanocomposites are capable of regulating the release of nutrients from the fertilizer capsule. Further, the activities of nanoparticles were closely monitored using Transmission Electron Microscope (TEM) and Scanning Electron Microscope (SEM). A planted nanocomposites consisting of N, P, K and micronutrients along with mannose and aminoacids have shown to increase the uptake and utilization of nutrients by grain crops (Jinghua, 2004). In order to regulate the release of nutrients from the fertilizers nano-zeolites are known to be used and found effective in enhancing nutrient use efficiencies by crops. Reduction of size through top down approaches (ball milling) appears to regulate the release of nutrients with or without surface modifications with suitable surfactants. Nutrient use efficiencies of nitrogen (Manikandan and Subramanian, 2013; 2014), phosphorous (Bansiwal *et al.*, 2006), potassium (Subramanian and Sharmila, 2012a), sulfur (Selva Preetha *et al.*, 2014a,b) and Zn (Subramanian and Sharmila, 2012b) are reported to be enhanced by such manipulations. These reports tended to indicate that a nanocomposite can be developed in order to supply all required essential elements by fortifying the nano-zeolites that can facilitate in achieving balanced crop nutrition and sustained farm productivity.

Smart Delivery Systems

Smart delivery system in agriculture is yet to be investigated. However, foliar feeding of nutrients and seed coating of agricultural inputs are considered "smart" delivery system as this technique will deliver the inputs directly to the plants and the site of action. But, the nano-based smart delivery systems should go beyond the boundaries of foliar feeding. The smart delivery system should consider the factors or combinations of factors such as time controlled, specifically targeted, highly controlled, remotely regulated/preprogrammed and multifunctional characteristics to avoid biological barriers for successful targeted release (Roco, 2003). These nano-formulations of various agrochemicals proved to be advantageous than conventional formulations applied to crops by spraying and broadcasting. According to Boehm *et al.* (2003), in conventional method of applying agro-chemicals, only a very low concentration of active ingredient, which is much below the minimum effective required concentration, has reached the target site of crops due to problems such as leaching, degradation by photolysis, hydrolysis and bio-instability in atmosphere. This in turn warrants the need of repeated application for the effective control of pests or weeds resulting in higher cost of cultivation and environmental pollution (Remya *et al.*, 2010).

Mode of Entry

Plant cell wall acts as a barrier for easy entry of any external agents including nanoparticles into the plant cells. The sieving properties are determined by pore diameter of cell wall ranging from 5 to 20 nm (Fleischer *et al., 1999*). Hence, only nanoparticles with diameter less than the pore diameter of the cell wall could easily pass through and reach the plasma membrane. There is also a chance for enlargement of pores or induction of new cell wall pores upon interaction with engineered nanoparticles which in turn enhance nanoparticle uptake. Further internalization occurs during endocytosis with the help of a cavity like structure that form around the nanoparticles by plasma membrane. They may also cross the membrane using embedded transport carrier proteins or through ion channels. In the cytoplasm, the nanoparticles are applied on leaf surfaces, they enter through the stomatal openings or through the base of the trichomes and then translocated to various tissues. However, accumulation of nanoparticles on photosynthetic surface causes foliar heating, which results in alterations of gas exchange due to stomatal obstructions that produce changes in various physiological and cellular functions of plants (Fernandez and Eichert, 2009).

Effects of Magnetic Nanoparticles

The main advantage of using magnetic nanoparticles is that they allow a very specific localization of the particles to release their load, which is of great interest in the study of nano-particulate delivery for plants. A few works have been reported regarding the uptake, translocation and specific localization of magnetic nano-particles (less than 50 nm) in pumpkin plants. Magnetization signals of various strengths were observed from different portions (ranging from roots to leaves) of the treated plants which clearly indicates the successful translocation of nanoparticles in the entire plant system irrespective of the area of application. No toxicity on plant growth had been detected thus suggesting the safe use of such nanoparticles for nano-particulate delivery in plants. Magnetic nanoparticles could be used as capping agents for porous carriers of various chemicals so that their specific localization and uncapping process has been carried out using external magnets, hence releasing the payload at target site. Such methods are possible for specific treatments in fruit trees or high-input crops under greenhouse conditions since it is easy to provide an external magnetic exposure that triggers the payload release (Zhu *et al.*, 2008; Corredor *et al.*, 2009).

Effects of Zinc Based Nanoparticles

The use of metal based nanoparticles like ZnO for increased permeability and creation of new holes in bacterial cell wall has paved way for the use of such nanoparticles for studying their cell internalization and further translocation in plants. Recently, in Tamil Nadu Agricultural University, Coimbatore, Natarajan *et al.* (2011) have shown that the accelerated aged pulses seeds treated with ZnO nano-particles had 9 times higher germination (90 per cent) as against the untreated seeds (10 per cent). Seeds that are stored for a long time generate reactive oxygen species (ROS) that are being quenched by the action of donating electrons from ZnO nano-particles. This technology is very appropriate for rainfed agriculture. On the other hand, seed germination and root growth study of zucchini seeds in hydroponic solution containing ZnO nanoparticles showed no negative effects whereas the seed germination of rye grass and corn was inhibited by nanoscale zinc (nano-Zn, 35 nm) and zinc oxide (nano-ZnO, 15–25 nm) respectively (Lin and

Xing, 2007; Stampoulis *et al.*, 2009). These articles suggest that nano-particles possess beneficial effects only when it is employed with caution.

Conclusion

The smart delivery systems are well established in medical sciences which made breakthroughs in finding solution to serious human diseases without associated side effects. The data from medical sciences serve as guiding tools that can be exploited in agricultural production systems. In this book chapter, the literature review made has clearly suggest that there is an abundance of scope to exploit smart delivery systems in agriculture which facilitate enhanced use efficiency of inputs besides environmental protection. It is time that agricultural scientists should undertake research in the fascinating field of nano-based smart delivery systems so as to achieve the targeted delivery of inputs that enhances the crop productivity of crops with minimal use agri-inputs.

References

Abraçado L G, Esquivel D M S, Alves O C and Wajnberg E 2005. Magnetic material in head, thorax, and abdomen of *Solenopsis substituta* ants: a ferromagnetic resonance study. *Journal of Magnetic Resonance* **175**: 309-316.

Bhagat D, Samanta S K and Bhattacharya S 2013. Efficient management of fruit pests by pheromone nanogels. *Scientific Reports*: 3.

Bansiwal A K, Rayalu S S, Labhasetwar N K, Juwarkar A A and Devotta S 2006. Surfactant-Modified Zeolite as a Slow Release Fertilizer for Phosphorus. *Journal of Agriculture and Food Chemistry* **54**: 4773-4779.

Barik T K, Sahu B and Swain V 2008. Nanosilica from medicine to pest control. *Parasitology Research* **103**: 253-258.

Boehm A L, Martinon I, Zerrouk R, Rump E and Fessi H 2003. Nanoprecipitation technique for the encapsulation of agrochemical active ingredients. *Journal of Microencapsulation* **20**: 433-441.

Corradini E, Moura, M R and Mattoso L H C 2010. A preliminary study of the incorporation of NPK fertilizer into chitosan nanoparticles Express. *Polymer Letters* **4**: 509-515.

Corredor E, Testillano P S, Coronado M, Gonzalez-Meledni P, Fernandez-Pacheco R, Marquina C, Ibarra M R, Rubiales D, Perez-de-Luque A and Risueno M C 2009. Nanoparticle penetration and transport in living pumpkin plants: *in situ* subcellular identification. *Plant Biology* **9**: 45.

DeRosa M C, Monrea l C, Schnitzer M, Walsh R and Sultan Y 2010. Nanotechnology in fertilizers. *Nature Nanotechnology* **5**: 1-2.

Douglas M Light and John J Beck 2010. Characterization of Microencapsulated pear Ester, (2E, 4Z)- Ethyl-2, 4-decadienoate, a kairomonal Spray Adjuvant against Neonate Codling Moth Larvae. *Journal of Agriculture and Food Chemistry* **58**: 7838-7845.

El-Jaick L J, Acosta-Avalos D, de S E D M, Wajnberg E and Linhares M P 2001. Electron paramagnetic resonance study of honeybee *Apis mellifera* abdomens. *European Biophysics Journal* **29**: 579-86.

Esquivel D, Wajnberg E, Nascimento F, Pinho M, Barros H and Eizemberg R 2007. Do geomagnetic storms change the behaviour of the stingless bee guiruc'/u (Schwarziana quadric-punctata). *Naturwissenschaften* **94**: 139-142.

Fernandez V and Eichert T 2009. Uptake of hydrophilic solutes through plant leaves: current state of knowledge and perspectives of foliar fertilization. *Critical Reviews in Plant Sciences* **28**: 36-68.

Fleischer,M A, O Neill R and Ehwald, R 1999. The pore size of non-graminaceous plant cell wall is rapidly decreased by borate ester cross-linking of the pectic polysaccharide rhamnogalacturon II. *Plant Physiology* **121**: 829-838.

Franca E F, Leite F L, Cunha R A, Oliveira J O N and Freitas L G 2011. Designing an enzyme-based nanobiosensor using molecular modeling techniques. *Physical Chemistry Chemical Physics* **13**: 8894-8899.

Frederiksen H K, Kristensen H G and Pedersen M 2003. Solid lipid microparticle formulations of the pyrethroid gamma-cyhalothrin-incompatibility of the lipid and the pyrethroid and biological properties of the formulations. *Journal of Control Release* **86**: 243-52.

Gan N, Yang X, Xie D, Wu Y, Wen W A 2010. Disposable organophosphorus pesticides enzyme biosensor based on magnetic composite nanoparticles modified screen printed carbon electrode. *Sensors* **10**: 625-638.

Gorb E and Gorb S 2009. Effects of surface topography and chemistry of Rumex obtusifolius leaves on the attachment of the beetle *Gastrophysa viridula*. *Entomologia Experimentalis Et Applicata* **130**: 222-228.

Guan H, Chi D, Yu J and Li X 2008. A novel photodegradable insecticide: preparation, characterization and properties evaluation of nano-Imidacloprid Pesticide. *Biochemistry and Physiology* **92**: 83-91.

Gunasekaran K 2011. Nano Pesticides. In Principles and Practices of Nano applications. *Shri Garuda Graphics* 139-152.

Guo Y R, Liu S H, Gui W J and Zhu G N 2009. Gold immunochromatographic assay for simultaneous detection of carbofuran and triazophos in water samples. *Anaytical Biochemistry* **389**: 32–39.

Jinghua G 2004. Synchrotron radiation, soft X-ray spectroscopy and nano-materials. *Journal of Nanotechnology* **1**: 193-225.

Kanimozhi V and Chinnamuthu C R 2012. Engineering Core/hallow Shell Nanomaterials to Load Herbicide Active Ingredient for Controlled Release. *Research Journal of Nanoscience and Nanotechnology* **2**: 58-69.

Kaushik A, Solanki P R, Ansarib A A, Malhotra B D and Ahmad S 2009. Iron oxide-chitosan hybrid nanobiocomposite based nucleic acid sensor for pyrethroid detection. *Biochemical Engineering* **46**: 132-140.

Knight A L, Larsen T E and Ketner K C 1988. Rainfastness of a Microencapsulated Sex Pheromone Formulation for Codling Moth (Lepidoptera: Tortricidae). *Journal of Economic Entomology* **97**: 1987-1992.

Lai F, Wissing S A, Müller R H and Fadda A M 2006. *Artemisia arborescens* L Essential Oil–Loaded Solid Lipid Nanoparticles for potential agricultural application: preparation and characterization. *AAPS Pharm SciTech* **7**: 1-9.

Li Z Z, Chen J F, Liu F, Liu AQ, Wang Q and Sun HY 2007. Study of UV-shielding properties of novel porous hollow silica nanoparticle carriers for avermectin. *Pest Management Science* **63**: 241-246.

Lin D and Xing B 2007. Phytotoxicity of nanoparticles: inhibition of seed germination and root growth. *Environmental Pollution* **150**: 243-250.

Lisa M, Chouhan R S, Vinayaka A C, Manonmani H K and Thakur M S 2009. Gold nanoparticles based dipstick immuno-assay for the rapid detection of dichlorodiphenyltrichloroethane: An organochlorine pesticide. *Biosens Bioelectron* **25**: 224-227.

Liu J, Wang F H, Wang L L, Xiao S Y, Tong C Y and Tang D Y 2008. Preparation of fluorescence starch-nanoparticle and its application as plant transgenic vehicle. *Journal of Central South University of Technology* **15**: 768-773.

Liu M, Liang R, Liu F and Niu A 2006. Synthesis of a slow release and superabsorbent nitrogen fertilizer and its properties. Polymers for *Advanced Technoogy* **17**: 430-438.

Liu Y, Yan L, Heiden P and Laks P 2001. Use of nanoparticles for controlled release of biocides in solid wood. *Journal of Applied Polymer Science* **79**: 458-465.

Manikandan A and Subramainan K S 2013. Urea Intercalated Biochar–a Slow Release Fertilizer Production and Characterisation. *Indian Journal of Science and Technology* **6**: 5579-5584.

Manikandan A and Subramainan K S 2014. Fabrication and characterisation of nanoporous zeolite based N fertilizer. *African Journal of Agricultural Research* **9**: 276-284.

Mihou A P, Michaelakis A, Krokos F D, Mazomenos B E and Coulasdouros E A 2007. Prolonged slow release of (Z)- 11- hexadecenyl acetate employing polyurea microcapsules. *Journal of Applied Entomology* **131**: 128-133.

Natarajan, N, Kalaivani S and Senthil Kumar S 2011. Nanotechnology applications in Seed Science. In Principles and practices of Nano applications. *Shri garuda Graphics* 130-138.

Nykypanchuk D, Maye M M, Gang O and Van D L D 2008. DNA-guided crystallization of colloidal nanoparticles. *Nature* **451**: 549-552.

Paknikar K M, Nagpal V, Pethkar A V and Rajwade J M 2005. Degradation of lindane from aqueous solutions using iron sulfide nanoparticles stabilized by biopolymers. *Science and Technology of Advanced Materials* **6**: 370-374.

Patolsky F, Zheng G and Lieber C M 2006. Nanowire sensors for medicine and life sciences. *Nanomedicine* 1: 51-65.

Perez-de-Luque A and Diego R 2009. Nanotechnology for parasitic plant control. *Pest Management Science* 65: 540-545.

Prakash V G 2013. Studies on pheromone dispensers.M.Sc. (Agri.) *Thesis,* Tamil Nadu Agricultural University, Coimbatore, India.

Priyanka Bansal, Kathrin Bubel, Seema Agarwal, and Andreas Greiner 2012. Water -Stable All-Biodegradable Microparticles in Nanofibers by Electrospinning of Aqueous Dispersions for Biotechnical Plant protection. Germany *Biomacromolecules.* pp. 439-444.

Rahman A, Seth D, Mukhopadhyaya S K, Brahmachary R L, Ulrichs C and Goswami A 2009. Surface functionalized amorphous nanosilica and microsilica with nanopores as promising tools in biomedicine. *Naturwissenschaften* 96: 31-38.

Rajamanickam K, Subaharan K, Srinivasan T and Paramagru P 2014. Evaluation of Nanomatrix Aggregation Pheromone (4 methyl-5 Nonanol) against Red Palm Weevil (*Rhynchophorus ferrugineus* Oliv.) in Coconut. **In:** *"National Symposium on ETEIPM" January,* 2014. pp.231.

Remya N, Saino N H, Baiju G N, Maekawa T, Yoshida Y and Sakthi K D 2010. Nanoparticle material delivery to plants. *Plant Science* 179: 154-163.

Roco M C 2003. Nanotechnology convergence with modern biology and medicine. *Current Opinion in Biotechnology* 14: 337-346.

Scrinis G and Lyons K N D 2007. *The emerging nano-corporate paradigm: Nanotechnology and the transformation of nature, food and agri-ffod systems. International Sociological Association. Internet Resource.* http://researchbank.rmit.edu.au/view/rmit: 8555.

Selva Preetha P, Subramanian K S and Sharmila Rahale C 2013. Sorption characteristics of nano zeolite based slow release sulphur fertilizer. *International Journal of Development Research (In Press).*

Selva Preetha P, Subramanian K S and Sharmila Rahale C 2013. Characterization of slow release of sulphur nutrient - a zeolite based nano-fertilizer. *International Journal of Development Research* 4: 225-228.

Singh M, Singh S, Prasad S and Gambhir I S 2008. Nanotechnology in medicine and antibacterial effect of silver nanoparticles. *Digest Journal of Nanomaterials and Biostructures* 3: 115-122.

Solgi M, Kafi M, Taghavi T S and Naderi R 2009. Essential oils and silver nanoparticles (SNP) as novel agents to extend vase-life of gerbera (Gerbera jamesonii cv. 'Dune') flowers. *Postharvest Biology and Technology* 53: 155-158.

Stampoulis D, Sinha S K and White J C 2009. Assay-dependent phytotoxicity of nanoparticles to plants. *Environmental Science and Technology* 43: 9473-9479.

Subaharam K, Eswarmoorthy, Pavan Kumar B V V S, Vibina Venugopal, Rajamanickam K, Srinivasan T, Chalapathi Rao N B V and Sanjeev Gurav 2014. Nanomatrix for delivery of semiochemicals of coconut Red Palm Weevil, *Rhychophorus ferrugineus*. In: *"National Symposium on ETEIPM"* January, 2014. pp. 215.

Subramanian K S and Sharmila Rahale C 2012. Ball Milled Nanosized Zeolite Loaded With Zinc Sulfate: A Putative Slow Release Zn Fertilizer. *International Journal of Innovative Horticulture* 1: 33-40.

Subramanian KS and Sharmila Rahale C 2012. Nano-fertilizer for smart delivery of nutrients. *Advances in Horticulture Biotechnology* 6: 307-316.

Subramanian KS and Sharmila Rahale C 2012. Slow release of potassium through nano-fertilizer. *International Journal of Nanotechnology and Applications* 6: 7-21.

Torney F, Trewyn B G, Lin V S, Wang K 2007. Mesoporous silica nanoparticles deliver DNA and chemicals into plants. *Nature Nanotechnology* 2: 295-300.

Vandergheynst J, Scher H, GuoHy and Schultz D 2007. Water-in-oil emulsions that improve the storage and delivery of the biolarvacide *Lagenidium giganteum*. *BioControl* 52: 207-229.

Vijayakumar P S, Abhilash O U, Khan B M and Prasad B L V 2010. Nanogold-loaded sharp-edged carbon bullets as plant-gene carriers. *Advanced Funcionalt Materials* 20: 2416-2423.

Wang H, Wang J, Choi D, Tang Z, Wu H and Lin Y 2009. EQCM immunoassay for phosphorylated acetylcholinesterase as a biomarker for organophosphate exposures based on selective zirconia adsorption and enzyme-catalytic precipitation. *Biosens Bioelectron* 24: 2377-2383.

Wilson M A, Tran N H, Milev A S, Kannangara G S K, Volk H and Lu G H M 2008. Nanomaterials in soils. *Geoderma* 146: 291-302.

Zhang X, Zhang J and Zhu K Y 2010. Chitosan/double-stranded RNA nanoparticle-mediated RNA interference to silence chitin synthase genes through larval feeding in the African malaria mosquito (*Anopheles gambiae*). *Insect Molecular Biology* 19: 683-693.

Zhu H, Han L, Xiao J Q and Jin Y 2008. Uptake, translocation and accumulation of manufactured iron oxide nanoparticles by pumpkin plants. *Journal of Environmental Monitoring* 10: 713-717.

Chapter 8

Policies Related to Biosecurity in Trade and Exchange of Germplasm in the Context of Plant Health Management

Kavita Gupta* and P C Agarwal

Division of Plant Quarantine, ICAR-National Bureau of Plant Genetic Resources, New Delhi – 110 012

*E-mail: kavita.gupta@icar.gov.in

ABSTRACT

Biosecurity policies play an important role in facilitating transboundary movement of plants/ plant products/plant genetic resources (PGR) by preventing introduction of pests in new geographical areas. International exchange of such material has contributed significantly towards widening our food basket, crop improvement and increased production. However, a number of pests have also moved across the countries along with these materials. With the coming of the two international Agreements of Convention on Biological Diversity (CBD) in 1992 and World Trade Organization (WTO) in 1995, international exchange of agricultural commodities has taken a different turn. Under WTO the aim is to promote trade by undertaking quarantine and influencing trade policies while CBD aims at protection and conservation of environment and biodiversity. A number of national regulations related to biosecurity or issues related to it have an impact on safe trade and exchange of PGR. Even the PGR being exchanged under the International Treaty on Plant Genetic Resources of Food and Agriculture are governed by the respective national biosecurity regulations of the nations. The Plant Quarantine (Regulation for Import into India) Order (2003) and its subsequent amendments indicate the government's eagerness to comply with the Sanitary and Phytosanitary Agreement of WTO. The National Seed Policy (2002), which encompasses quality assurance mechanisms and facilitates a vibrant seed industry has emphasized on the importance of excluding diseases and pests in the national seed production programmes. The Insecticides Act (1968), which registers new molecules, including botanical pesticides, takes care of their efficacy and safety to human beings and animals. The provisions under the Protection of Plant Variety and Farmers' Rights Act (2001) for protection of plants/environment and the Biological Diversity Act (2002) needs to develop a mechanism to restrict the movement of infected/infested material within the country. The Environment Protection Act (1986) needs to make suitable provisions for dealing with Invasive Alien Species, and exchange of genetically modified crops after effectively complying with the Cartagena Protocol on Biosafety of the CBD. Under such a fragmented national system there is a need to ensure that our national legislations are in harmony with each other while complying with the international norms. There is also a need to support research, training, capacity-building, networking and information sharing activities.

Keywords: Biosecurity, Policies, Regulation, Holistic, Plant pests.

Introduction

Biosecurity literally means safety of living things. Technically it is a strategic and integrated approach that encompasses the policy and regulatory frameworks to analyze and manage risks in the sectors of food safety, animal life and health, and plant life and health, including associated environmental risk (http://www.fao.org). Biosecurity covers the aspects of security related to introduction of plant pests, animal pests and diseases, and zoonoses, and also the introduction and release of genetically modified organisms (GMOs) and their products. Thus, it is a holistic concept that covers safe transboundary movement of commodities, the potential threat of pests and diseases in biowarfare and also the indigenous threat of emerging pests due to changing climate and cropping pattern, and biosafety issues related to use of transgenics. This paper primarily confines to the regulations impacting phytosanitary aspects of biosecurity with a focus on dangers from exotic pests, status of national policies and the need for developing a holistic biosecurity policy to address the issue.

Need for Biosecurity

The fast paced developments in the previous two decades have created conditions which pose a new level of risk to human, plant and environmental health. Globalization, frontier biological innovations such as genetic engineering and climate change have the potential to impact human and plant life and our environment in unprecedented ways. Further, the advent of WTO and the liberalization of global trade in agriculture since 1995 has opened new avenues for growth and diversification of agriculture, but has also brought in many new challenges. There is an increased risk of introduction of exotic pests, including weeds, in the country with the potential to cause serious economic loss. Advances in genetic engineering leading to the introduction and release of living modified organisms (LMOs) or their products (*e.g.* GMOs) require additional risk analysis and proper risk management. Climate change has the potential to alter the habitat of known pests and even cause introduction of new pests. Also, there is an ever increasing threat of bioterrorism. The recent emergence and spread of transboundary diseases such as the Ug_{99} stem rust of wheat pose new threats to human, animal and plant safety.

It is because of these factors that biosecurity has emerged as one of the most urgent issues facing countries requiring them to foster policies and develop technological capabilities to prevent, detect, and respond to such threats. Biosecurity covers food safety, zoonoses, the introduction of animal and plant diseases/pests, the introduction and release of LMOs or GMOs and the introduction and management of invasive alien species. Biosecurity is a strategic and integrated approach based on the recognition of critical linkages between various sectors and the need for harmonizing and integrating national biosecurity systems and controls to take advantage of synergies across sectors. This would increase the national capacity to protect human health, agricultural production and livelihood, safeguard the environment, and protect against uncertain technologies. In addition, it would equip countries to meet obligations for compliance to the international trade under the Agreement for Application of Sanitary and Phytosanitary Measures in food and agricultural products. The overarching goal of biosecurity is to prevent, control and manage risks to life and health of humans, animals, plants.

The introduction of pests into a new locality is brought about in various ways: (a) the host may be the carrier; (b) inert materials such as packing material may carry resting stages of the organism; (c) insect vectors and birds may transport it; (d) air currents may carry the pest to long distances or (e) there may be deliberate, illegal introductions to use them as bioweapons. The first two modes of distribution lend themselves to curtailment by quarantine measures. The next two, are by and large beyond human control and are a major limitation in the control of pest by exclusion. A different degree of alertness and preparedness is required for the last category. The global trade in agricultural commodities and transboundary movement of genetic resources of plants and animals has led to situations in the past which warranted legislative measures to regulate such trade/exchange.

Plant quarantine is the government endeavor enforced through legislative measures to regulate the introduction of planting materials, plant products, soil, living organisms, etc. in order to prevent inadvertent introduction of pests (including fungi, bacteria, viruses, nematodes, insects and weeds) harmful to the agriculture of a country/state/region, and if introduced, prevent their establishment and further spread. The devastating effects resulting from diseases and pests introduced along with international movement of planting material, agricultural produce and products are well documented (Khetarpal, 2004, Gupta *et al.*, 2005, Khetarpal *et al.*, 2009), which clearly demonstrate that introduction and establishment of quarantine pests including weeds into new areas can severely damage the crop production and economy of a region/country.

Plant Biosecurity Policies in India

After the Second World War, FAO convened an International Plant Protection Convention (IPPC) in 1951, to which India became a party in 1956 along with Australia, Sri Lanka, UK, Netherlands, Indonesia, Portugal and Vietnam. At present, there are 182 signatory members of the IPPC (as on 12 January 2016). The IPPC aims to develop international cooperation among various countries to prevent the introduction and spread of regulated pests along with international movement of plants and planting material (http://www.ippc.org) and requires that each country establishes a national plant protection organization to discharge the functions specified by it.

With the coming of the two international Agreements of Convention on Biological Diversity (CBD) in 1992 and World Trade Organization (WTO) in 1995, international exchange of agricultural commodities has taken a different turn. Under WTO, the aim is to promote trade by undertaking quarantine and influencing trade policies while CBD aims at protection and conservation of environment and biodiversity. India is a signatory and has ratified all three international instruments impacting transboundary movement of plants/planting material needs to comply with the requirements under each. A number of national regulations related to biosecurity or issues related to it have an impact on safe trade and exchange of PGR. Even the PGR being exchanged under the International Treaty on Plant Genetic Resources of Food and Agriculture are governed by the respective national biosecurity regulations of the nations.

The chronology of notification of Indian Acts and Policies related to plant biosecurity and the provisions in brief covered therein are given in Table 8.1.

Infrastructure

In India agriculture is in the State list of the Constitution and crop production, including plant protection from indigenous pests, is the responsibility of State Governments. The Central Government (Ministry of Agriculture, Department of Agriculture and Cooperation) supplements States' efforts in pest surveillance and management by disseminating innovative pest management technologies. Plant quarantine and locust control in the scheduled desert area, being inter-state and international subjects, are managed by the Government of India.

The Department of Agriculture and Cooperation (DAC) through its Directorate of Plant Protection, Quarantine and Storage (DPPQS) implements schemes of plant protection and quarantine through its Central Integrated Pest Management Centers (CIPMCs), Plant Quarantine Stations (PQS) at different airports, seaports and land frontiers, Locust Warning Organization circle offices and Central Insecticides Laboratory (CIL) and Regional Pesticides Testing Laboratories (RPTLs), Central Insecticide Board and Registration Committee (CIB and RC) and National Plant Protection Training Institute (NPPTI), which has been upgraded as the National Institute of Plant Health Management (NIPHM) in October, 2008. In all, two categories of materials being imported under the PQ Order, 2003: (a) bulk consignments for consumption and sowing/planting, and (b) samples of germplasm in small quantities for research purposes. The PQS under the DPPQS undertake quarantine processing and clearance of consignments of the first category (http://www.plantquarantineindia.org). The five major Plant Quarantine Stations have been modernized recently under an FAO-UNDP funded project.

Plant quarantine regulations are implemented under the Destructive Insects and Pests Act, 1914 and the Plant Quarantine (Regulation of Import into India) Order, 2003. Other regulations relevant to biosecurity are the Environment (Protection) Act, 1968, the Biological Diversity Act, 2002 and the Prevention of Food Adulteration

Table 8.1: Indian Regulations Related to Plant Biosecurity from the Turn of the Century

Year	Legislation	Key Features Related to Plant Protection
1914	Destructive Insects and Pests (DIP) Act	☆ Enacted as the first quarantine law in India after the British ordered compulsory fumigation of imported cotton bales to prevent Mexican cotton boll weevil (*Anthonomus grandis*) ordered in 1906 ☆ Prohibits or regulates the import into India or any part thereof or any specific place therein or any article or class of articles specified therein ☆ Prohibits or regulates the export from a State or the transport from one State to another State in India of any plants and plant materials, diseases or insects, likely to cause infection or infestation ☆ Authorizes the State Govt. to make rules for detention, inspection, disinfection/disinfestation or destruction of any pest or class of pests or of any article or class of articles for which center has issued notifications ☆ Revised and amended several times over the years with provisions for domestic quarantine of nine pests
1966	Seeds Act	☆ Act was passed by the parliament in 1966 and the Seed Rules were framed under it in 1968 ☆ Legislative measures promulgated to ensure high quality of seed in the market place ☆ Act provides for a system of notification of kinds or varieties of seeds ☆ Seed (Control) Order, 1983 seeks to control and regulate seed production and distribution ☆ Seed declared as an essential commodity under the Essential Commodities Act, 1955 ☆ Periodically revised with the changing national and international scenario
1968	Insecticides Act	☆ Regulates import, manufacture, sale, distribution and use of insecticides with a view to prevent risk to human beings or animals and promotes safety measures ☆ Registration committee registers insecticides after verifying claims regarding efficacy and safety
1984	Plants Fruits and Seeds (Regulation for Import into India) Order	☆ No consignment would be imported into India without a valid import permit issued by the competent authority for a. bulk consignments – the Plant Protection Advisor to the Govt. of India b. importing germplasm of agri-horticultural crops including transgenic material for research- the Director, NBPGR c. forest plants- the Forest Research Institute, Dehradun d. remaining plants of economic and general interest- the Botanical Survey of India, Kolkata
1989	Revised PFS Order	☆ No consignment can be imported unless accompanied by an official Phytosanitary Certificate issued by an official agency of the exporting country ☆ Seeds/planting materials requiring isolation growing under detention, to be grown in an approved post-entry quarantine facility ☆ Import of soil, earth, sand, compost, plant debris accompanying seeds/planting materials is not permitted. Hay, straw or any other material of plant origin would not be used as packing material ☆ Special conditions for import of plants, seeds for sowing, planting and consumption have also been mentioned under Schedule II (Clause 4) of the Order

Contd...

Table 8.1–*Contd...*

Year	Legislation	Key Features Related to Plant Protection
1986	Environment Protection Act	☆ Aims to protect and improve quality of environment ☆ Provides for management and handling of hazardous wastes, use, import/export and hazardous micro-organisms, genetically engineered organisms or cells ☆ Both LMOs and invasive alien species are covered under the EPA; but, it does not state in clear terms the modality for restriction
1988	New Policy on Seed Development	☆ Came into force with an objective to provide the Indian farmers with the best genetic material available anywhere in the world to increase agricultural productivity, farm income and export earnings ☆ Encourages an export-oriented horticultural industry the government wanted to provide best seed or other planting material like bulbs, cuttings, saplings and bud woods freely to the farming community for export promotion ☆ Aimed at liberalization of imports along with streamlining of plant quarantine procedures and encouragement to domestic seed industry through incentives
2001	Protection of Plant Varieties and Farmers Rights Act	☆ Has set up a Plant Varieties and Farmers' Rights Protection Authority ☆ Allows the registration of new plant varieties within a specific list of genera and species, as well as farmers' varieties ☆ Accords specific privileges to researchers/breeders while respecting quarantine regulations

2002	Biological Diversity Act	☆ Addresses access to genetic resources, associated knowledge and equitable sharing of benefits arising there from to the country
		☆ To safeguard the interest of the people of India few exceptions are:
		  a. Use by *vaids* and *hakims*
		  b. Free access to the Indian citizens to use within the country for research
		☆ Collaborative research subject to policy guidelines and approval

2002	National Seeds Policy	☆ The Seeds Act, 1966, Seeds Control Order and NPSD, 1988 promote and regulate the seed industry ☆ Encompasses quality assurance mechanisms along with facilitation of seed industry ☆ Thrust areas include quarantine of imported seeds and planting material and compliance to biosafety ☆ A specified quantity of imported seeds to be sent to Gene Bank, NBPGR
2003	Plant Quarantine (Regulation for Import into India) Order	☆ Notified in compliance to the Sanitary and Phytosanitary Agreement of WTO ☆ Pest risk analysis made a precondition for all imports ☆ Gives various schedules for import of various plants and planting materials ☆ Quarantine fee structure also rationalized ☆ The earlier PFS order has been repeated under this order

Contd...

Table 8.1–Contd...

Year	Legislation	Key Features Related to Plant Protection
2009-14	Foreign Trade Policy	☆ Provides policy measures for enhanced market access across the world and diversification of export markets. ☆ To increase India's percentage share in global trade special initiatives identified for market diversification, technological upgradation, support to stakeholders in Agriculture among other sectors ☆ Import of restricted items allowed such as pesticides are permitted under Advanced Authorization for agro exports. ☆ Certain specified flowers, fruits and vegetables are entitled to a special duty credit
2010	Seed Bill	☆ More emphasis give to seed quality including health aspects ☆ Any traded seed should be identified in terms of the variety to which it belongs, meet the minimum prescribed standards for germination, genetic and physical purity, maximum standard in seed health and an acceptable level of agronomic performance ☆ Certification of seed standards is mandatory and done by accredited Seed Testing Laboratories
2013	Agricultural Biosecurity Bill	☆ Provides for establishment of an Authority for prevention, control eradication and management of pests and diseases of plants, animal and unwanted organisms for ensuring agricultural biosecurity. ☆ Most International obligations of India for facilitating imports and exports of plants, plant products, animals, animal products, aquatic organisms and regulation of agriculturally important microorganisms. ☆ Provisions for effective domestic quarantine incorporated. ☆ Regulates introduction of new or beneficial organisms in the country. ☆ Regulates impact of transgenic material with respect to sanitary and phytosanitary matters. ☆ Empower the quarantine officers for search and seizure of material.

Act, 1955. Plant quarantine regulations aim to prevent introduction of exotic pests, diseases and weeds into India, during the import of agricultural commodities, in accordance with the WTO-SPS Agreement. Phytosanitary Certificates (PSCs) are issued for exports as per the International Plant Protection Convention (IPPC) 1951 of the FAO and post-entry quarantine (PEQ) inspections undertaken for imports. National Bureau of Plant Genetic Resources (NBPGR) undertakes the quarantine processing of all plant germplasm and transgenic planting material under exchange for which it has well- equipped laboratories, green house complex and a Containment Facility has also been established for processing transgenics (Khetarpal *et al.*, 2006). NBPGR also provides the necessary back-up research. NBPGR also has a well-equipped quarantine station at Hyderabad, which mainly deals with the export samples of International Crops Research Institute for Semi-arid Tropics (Chakrabarty *et al.*, 2005).

Gaps in National Plant Biosecurity System Requiring Strengthening

I Legislation

The Destructive Insects and Pests Act (DIP Act), 1914 and the PQ Order 2003 cover import of plants/plant products based on risk assessments and management there is a need to strengthen term with respect to the following:

☆ The DIP Act is an old legislation and subsidiary to the Sea and Customs Act, which does not give direct powers to the Plant Quarantine Officers to deport or destroy or confiscate the consignment or lodge complaints under the Indian Penal Code; powers for seizures/detention or forfeiture/ confiscation of infected/infested material need to be incorporated; punishment or penalty on the importers or custom house clearing agents or other defaulters for violation of the Acts needs to be introduced and the provisions for regulating or prohibiting plants need to be strengthened. Also, the Act contains several definitions which need revision and updation.

☆ Although the DIP Act has the provisions for domestic plant quarantine, there are no supporting clauses for its enforcement resulting in ineffective enforcement.

The enactment of the Agricultural Biosecurity Bill 2013 would take care of the above issues.

II Infrastructure

Establishment of new plant quarantine stations with adequate facilities, manpower to regulate imports through notified points of entry, pest monitoring and control

Government of India has notified more than 130 International Entry Points which are being managed by only 57 functional Plant Quarantine Stations. There is an urgent need for new PQ stations with good infrastructure and trained technical personnel. The existing PQ Stations are required to be strengthened with additional human and infrastructural resources in view of the rapid increase in the quantum of international trade.

Further, strengthening of technical human resources is required for pest monitoring and surveillance, pesticide registration and testing systems, etc. The five major Plant Quarantine stations need to be strengthened/established to act as referral centres for rapid and accurate diagnosis of plant pests/pathogens.

III Capacity Building and Modernization

Development of an Integrated Information Management System

There is a need to improve the information management system used by the plant and animal quarantine services to meet organizational and client needs by way of free exchange of information. The Plant Quarantine Information System recently launched by the DPPQS, is a good step in this direction. The PQIS accessible from http://plantquarantineindia.nic.in, provides an efficient and effective service for the stakeholders such as importers, exporters, individuals and the Government. It facilitates importers to apply online for Import Permit, Import Release Order and Exporters to apply online for Phytosanitary Certificate. Exporters and importers can view the status of their application online and access history of their applications during the entire life cycle of the application. This will help in bringing transparency in functioning.

Establishment of Biosecurity and Trade Unit

There is an urgent need to establish a Biosecurity and Trade Unit with various Cells namely;

☆ *Sanitary and phytosanitary cell* dealing with requirements for import of plants and planting material in the SPS-WTO regime, scientific justification to facilitate pest/disease free trade in agricultural commodities, fulfill the obligations of the SPS Enquiry Point, examine the WTO-SPS notifications from other countries, preparing India's SPS notifications for WTO, preparing India's market access claims, bilateral agreements with countries etc (Khetarpal and Gupta, 2002).

☆ *Risk analysis unit for import and export and market access cell* to undertake science based pest risk analysis for the application of SPS measures consistent with the WTO-SPS Agreement and IPPC international standards on pest risk analysis (Gupta and Khetarpal, 2004). The unit would also organize trainings for capacity building in plant quarantine, develop national standards and harmonize with the international standards.

☆ *Integrated pest surveillance and rapid response cell* to ensure early detection of introduced pests, to provide reliable data for Pest Risk Analysis and monitoring pest status in an area to support market access through "pest-free areas". There is a need to conduct regular surveys to watch the introduction, establishment and spread of pests and diseases. In case of any eventuality, a Rapid Response Team would be constituted to check and control the spread of the disease

☆ *Centralized biosystematics cell* needs to be established for identification of pests/diseases faster to enhance the decision making by the operational staff at port of entries. The unit will comprise experts in the fields of entomology, plant pathology including nematodes and weed science with supporting staff for identification of the pests/diseases. The ICAR-DARE bureaux for microorganisms and insects (arthropods) will be utilized as resource base for this purpose. They would use the latest tools such as molecular techniques, digitized keys and remote microscopy for speedy and accurate identification of pests.

☆ *Emergency action and biosecurity disaster management cell* to be developed as an integrated unit with specialized experts for undertaking biosecurity risk assessment, draw out Emergency Action Plan to combat pest menace in the event of any pest epidemic by integrating the various control techniques so as to timely manage the pest incidence and avoid crop loss/yield. It could have four wings dealing with plants, farm animals, living aquatic resources and agriculturally important micro-organisms.

☆ *Human resource development cell* for regular trainings on sanitary and phytosanitary issues for identification of pest diagnostics, sampling, international standards/guidelines etc. is required. This Cell would also organize trainings, seminars, workshops etc. on sanitary and phytosanitary issues and standards along with related standard operating procedures, at domestic and international level. Besides, the Cell would also sensitize policy makers/implementers, administrators, politicians, stakeholders (village farmers, farmers owning large organized farms), general public to the importance of food safety, good agricultural practices, plant quarantine requirements through seminars, symposia, workshops and mass media programmes is required.

IV Strengthening Research Back-up

☆ Classification of the threat agents including insects, mites, microorganisms and bio-weapons challenging the biosecurity of agriculture based on perceived risk levels.

☆ Increased use of radiation and other frontier techniques as effective mitigation treatment.

☆ Development of user-friendly serological/molecular diagnostic protocols/kits for prognostic detection of exotic pests and their variants and also for the detection of GMOs/LMOs.

☆ Development of digitized biosystematic keys for identification of pests.

☆ Epidemiological studies including survey and surveillance of diseases/pests to prepare database on endemic pests, identify pest free areas and target IPM for reducing threats.

☆ Developing models for risk analysis for exotic pests, diseases, invasive weeds and genes.

☆ Developing standard operating procedures through relevant Handbooks/ Manuals for survey and surveillance and monitoring major diseases/pests including invasive weeds.

☆ Studies on epidemiology of economically important pests and diseases vis-à-vis climate change.

☆ Studies on factors affecting likelihood of survival of pests under different conditions of transport, modes of dispersal, distribution of hosts/alternate hosts at destination, potential for establishment, reproductive strategy and method of pest survival, potential vectors and natural enemies of the pest in the area.

☆ Simulated evaluation of mitigation options to deal with epidemics/ pandemics.

☆ Need for special focus on management of indigenous minor diseases, pests and invasive weeds with a potential to impact food security, environment including biodiversity, and trade.

☆ Development of national dynamic biosecurity database of pest, diseases, invasive weeds of plants/animals/poultry/fish and their management.

Need for a Harmonized and Integrated Approach to Agricultural Biosecurity

Internationally, the Agreement on the Application of Sanitary and Phytosanitary Measures of the WTO, governs SPS measures in relation to international trade. The Codex Alimentarius Commission (Codex), the IPPC and the Office International des Epizooties (OIE) provide international standards for food safety, plant health, and animal health, respectively. Further, the Cartagena Protocol of Convention on Biological Diversity (CBD) applies to the transboundary movement, transit, handling and use of Living Modified Organisms (LMOs). Guidelines on the management of invasive alien species have been developed under the SBSTTA (Subsidiary Body on Scientific, Technical and Technological Advice) of CBD (Rana *et al.*, 2004, Gupta *et al.*, 2009). This group of international agreements, organizations and programmes are part of a loose international framework for biosecurity, and reflects the sectorial approach to regulate this area.

FAO has recognized the growing importance of biosecurity, and made it one of its sixteen Priority Areas for Inter-disciplinary Action. Biosecurity was also included in the Medium Term Plan which aims at "promoting, developing and reinforcing policy and regulatory frameworks for food, agriculture, fisheries and forestry" (http://www.fao.org.COAG/2003/9.htm).

Models to rationalize regulatory functions among sectors in the quest for improved effectiveness and efficiency have appeared in a number of countries. New Zealand has a Biosecurity Act since 1993 and a Biosecurity Minister and Council since 1999. In US, agricultural biosecurity is looked after by the Animal and Plant Health Inspection Service (APHIS) headed by an administrator under the US Department of Agriculture (USDA). In Australia, biosecurity is the charge of the Department of Agriculture, Fisheries and Forestry (DAFF) looking after plant and animal biosecurity policy development and risk management measures and international trade. The quarantine machinery of New Zealand, USA and Australia have been totally transformed and upgraded in their own ways to meet the challenges of their national biosecurity and safe trade.

Presently in India, agricultural biosecurity is managed on a sectoral basis through the development and implementation of separate policy and legislative frameworks (*e.g.* for animal and plant life and health). Sectoral agencies organize their work without much attention to the other sectors and no attention is paid to the interdisciplinary nature of biosecurity. In a modern national system, there is need for a more harmonized and integrated approach, with competent authorities responsible for different sectors of agricultural biosecurity working together towards common goals. Sectoral policies, laws and regulations can be harmonized to avoid contradictions, overlaps and gaps. This encompasses the joint setting of biosecurity priorities and allocation of resources, joint planning of activities, and integrated systems for monitoring and review of outcomes. An integrated agricultural biosecurity system would present a single interface to exporters/importers and allows for sharing of resources among the sectoral agencies.

Accordingly, an Agricultural Biosecurity Bill 2013 has been drafted by the Ministry of Agriculture and Farmers Welfare wherein a national mechanism under whose umbrella all the related issues are dealt with in a comprehensive manner to achieve national agricultural biosecurity, protect animal and plant life and health, environment and food safety (Figure 8.1). There is a need for synergy of expertise from various organizations under the Ministries of Agriculture and Farmers Welfare,

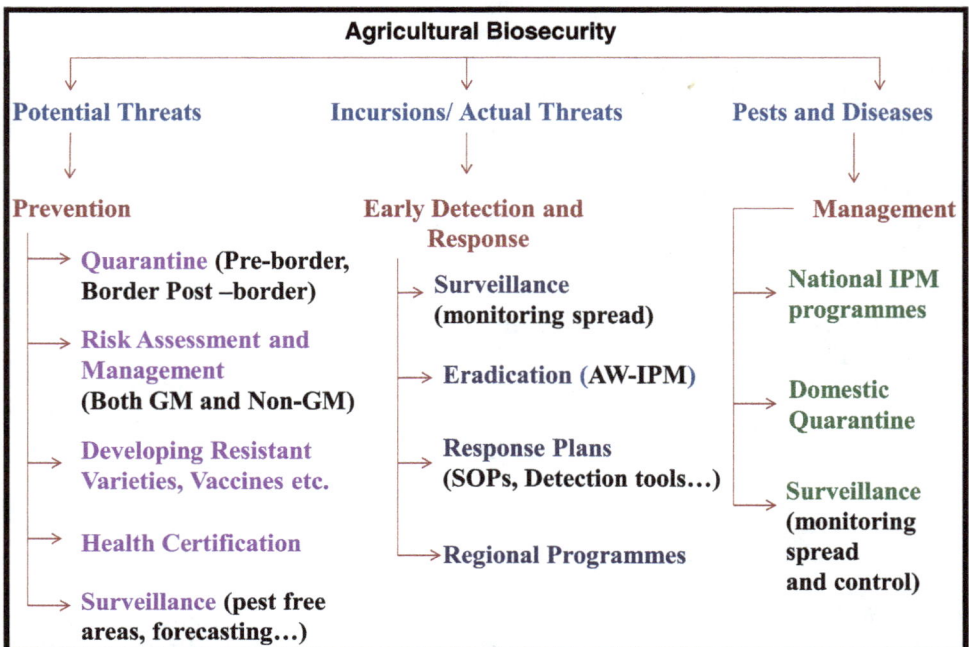

Figure 8.1: Strategies for Holistic Biosecurity.

Environment and Forests, Commerce and Industry, Rural Development, Health and Family Welfare, Science and Technology, Home Affairs, Defence, National Disaster Management Authority. Integration of the system has been done at the top level for taking prompt decisions and conveying firm policy guidelines for managing risk of entry, establishment or spread of pests and diseases and undertaking action plans to regulate them effectively. The integrated system will facilitate (a) timely handling of calamities, (b) appropriate utilization of resources, and (c) establishment of a systems approach to achieve desired results. Integration would imply sharing of national facilities and expertise across different disciplines and institutions like ICAR, CSIR, and SAUs etc. In the light of the various issues as highlighted, this bill will address the problem of biosecurity in the country holistically.

A holistic model proposed a few years ago depicting the various components of biosecurity including biowarfare, production, trade, environment (biodiversity and biosafety) is given in Figure 8.2 (Khetarpal and Gupta, 2007). It also shows the regulatory mechanism including various committees in place at national level [Export Import Policy (EXIM), EPA, Biological Diversity Act (BDA), Protection of Plant Varieties and Farmers Rights Act (PPVFRA), RCGM and GEAC] for implementation of the international agreements [Biological Weapons Convention, CBD, Cartagena Protocol (CP) and SPS/WTO]. The various stakeholder ministries involved are also depicted in the figure. A coordinated effort and synergy is all what is required to have an effective biosecurity system in the country.

For a biosecured India, what we need is a synergy of expertise from various organizations under the Ministries of agriculture, food, science and technology, home affairs, commerce and industry and defence. Presently, at the national level,

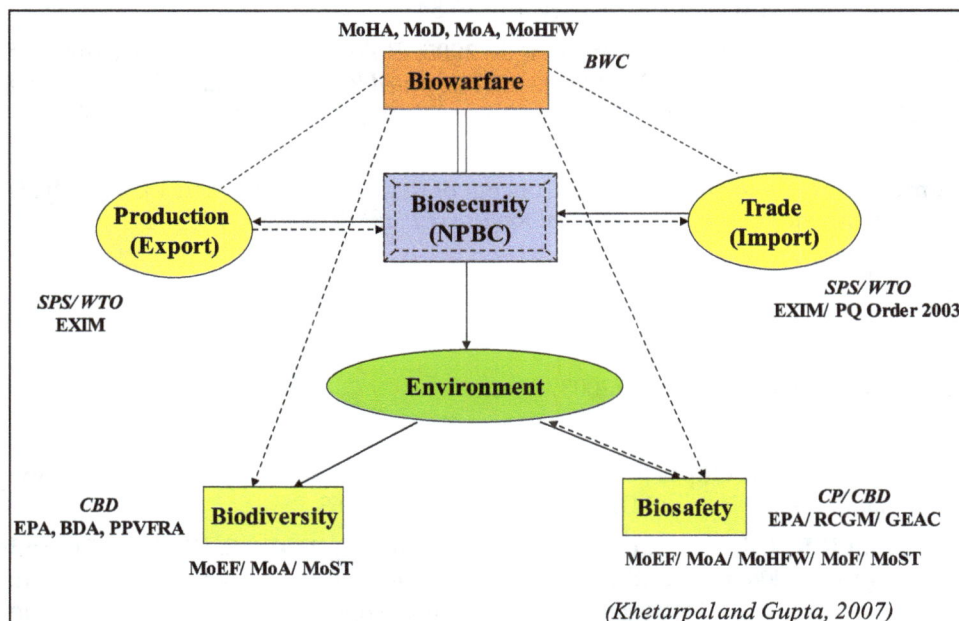

(Khetarpal and Gupta, 2007)

Figure 8.2: A Holistic Model for National Plant Biosecurity System.

efforts being made for the development of a coherent biosecurity strategy for the country by the formulation of a comprehensive Agricultural Biosecurity Bill, 2013 is in progress. DAC has initiated the establishment of a National Agricultural Biosecurity System with a National Agricultural Biosecurity Centre at DPPQS and a National Agricultural Biosecurity Network with the stakeholders from various ministries, ICAR, SAUs and state departments (Swaminathan, 2008). Also, the drafting of the Biosecurity Bill in 2013 by the DAC to address the issue of national biosecurity in a holistic manner are steps in the right direction.

Acknowledgements

The authors sincerely acknowledge the contributions of Dr R K Khetarpal, Former Head, Plant Quarantine Division, who initiated work on policy issues related to biosecurity at NBPGR and has a major role in shaping our understanding of the subject.

References

Chakrabarty S K, Anitha K, Girish A G, Sarath Babu B, Prasada Rao R D V J, Varaprasad K S, Khetarpal R K and Thakur R P 2005. *Germplasm Exchange and Plant Quarantine of ICRISAT Mandate Crops*. ICRISAT, Patencheru, Hyderabad 72p.

Gupta K, Chalam V C, Dev U and Khetarpal R K 2009. Global Information on Invasive Alien Species. In: *Invasive Alien Species- A Threat to Biodiversity*. (eds. Sharma S K, Khetarpal, R K, Gupta, K, Lal, A and Venkatraman, K) Indian Council of Agricultural Research, Ministry of Environment and Forests and National Biodiversity Authority. pp. 39-54

Gupta K, Gaur A and Khetarpal R K 2005. Role of Regulatory Measures in Controlling Spread of Plant Pests, In: *Integrated Pest Management: Principles and Application Vol. I*. (eds. Singh A, Sharma O P and Garg D K) CBS Publishers and Distributors, New Delhi pp. 48-69.

Gupta K and Khetarpal R K 2004. Concept of Regulated Pests, their Risk Analysis and the Indian Scenario. *Annual Review of Plant Pathology* 3: 409- 441.

http: //plantquarantineindia.nic.in

http: //www.plantquarantineindia.org

http: //www.fao.org

http: //www.fao.org.COAG/2003/9.htm

http: //www.ippc.org

Khetarpal R K and Gupta K 2007. Plant Biosecurity in India- Status and Strategy. *Asian Biotechnology and Development Review* 9: 39-63.

Khetarpal R K, Lal A, Varaprasad K S, Agarwal P C, Bhalla S, Chalam V C and Gupta K 2006. Quarantine for SafeExchange of Plant Genetic Resources. In: *100 years of PGR Management in India* (eds. Singh A K, Saxena S, Srinivasan

K and Dhillon B S) National Bureau of Plant Genetic Resources, New Delhi, India pp. 83-108.

Khetarpal R K 2004. A critical Appraisal of Seed Health Certification and Transboundary Movement of Seeds under WTO Regime. *Indian Phytopathology* **57**: 408-427.

Khetarpal R K and Gupta K 2002. Implications of Sanitary and Phytosanitary Agreement of WTO on Plant Protection in India. *Annual Review of Plant Pathology* **1**: 1-26.

Khetarpal R K, Balaraman N, Bandopadhyay S K, Gupta K, Jain R K, Kumar J, Mukhopadhyay A N and Vijayan K K 2009. Status of Agricultural Biosecurity. In: *State of Indian Agriculture*. National Academy of Agricultural Sciences, NAS Complex, New Delhi. pp. 151- 174.

Rana R S, Dhillon B S and Khetarpal R K 2004. Invasive Alien Species: The Indian Scene. *Indian Journal of Plant Genetic Resources* **16**: 190-213.

Swaminathan M S 2008. Preparedness for ensuring Biosecurity. In: *Biosecurity* (eds Shetty P K, Parida Ajay and Swaminathan M S) National Institute of Advanced Studies, Bangalore and M S Swaminathan Research Foundation Chennai, India. pp. 1- 9.

Chapter 9

Pesticide Residues in Food: An Indian Perspective

K K Sharma[1] and Sreenivasa Rao Cherukuri[2]

[1]*Network Coordinator, All India Network Project on Pesticide Residues,*
ICAR-Indian Agricultural Research Institute, New Delhi – 110 012
[2]*All India Network Project on Pesticide Residues,*
Professor Jayashankar Telangana State Agricultural University,
Rajendranagar, Hyderabad – 500 030, Telangana State

ABSTRACT

Pesticides as a plant protection tool helped in increased food production, but on the other side of the success, the problems such as presence of pesticide residues in food and feed, environmental pollution, pest resistance, pest resurgence, outbreak of secondary pests, killing of non-targets including natural enemies, pollinators etc were also noticed, and out of all these, the presence of pesticide residues in foods is a cause of serious concern by the general public. In India, pesticide residues in foods are regulated under Food Safety and Standards Act, 2005. In order to estimate the residue contamination in foods at both state and national level, on the recommendation of Joint Parliamentary Committee (JPC) on Pesticide Residues during 2003, the Department of Agriculture and Cooperation (DAC), Ministry of Agriculture has started the central sector scheme, "Monitoring of Pesticide Residues at National Level" during 2005-06 with participation of >32 laboratories across the nation. Monitoring studies conducted during 2006-2012, indicate that around 15 per cent of different vegetables are detected with one/more pesticide residues, and the matter of concern is that in about 3 per cent vegetables pesticide residues were detected above legal limits i.e. Maximum Residue Limits. However, in case of fruits, out of 9 per cent pesticide residue detected samples, about 0.9 per cent samples were contaminated with pesticide residues above MRLs. These data help focus governmental interventions such as Integrated Pest Management and other suitable measures in regions of high prevalence of pesticide residues.

Introduction

Food and Agricultural Organization (FAO) defines pesticide as any substance or mixture of substances intended for preventing, destroying, or controlling any pest, including vectors of human or animal diseases, unwanted species of plants or animals causing harm during, or otherwise interfering with the production, processing, storage, transport, or marketing of food, agricultural commodities, wood and wood products, or animal feedstuffs, which may be administered to animals for the control of insects, arachnids, or other pests in or on their bodies. The term includes substances intended for use as a plant-growth regulator, defoliant, desiccant, fruit-thinning agent, or an agent for preventing the premature fall of fruit, and substances applied to crops either before or after harvest to prevent deterioration during storage or transport. The term excludes fertilizers, plant and animal nutrients, food additives, and animal drugs.

Pesticides have played an important role in Indian agriculture during the green revolution for its food security programme and will remain indispensable in future also. The various pests reduce the food production by about 30 per cent, worth thousands crores of rupees, in the field, storage, transportation etc without the pesticide umbrella. Pesticides are used in agriculture for three major purposes *viz.,* i) to protect the crops from pest attack, ii) to produce a crop of higher quality and iii) to reduce the input of labour and energy as in case of herbicide usage. In India, 240 insecticides/pesticides were registered under section 9(3) of the Insecticide Act, 1968 for use in the country as on 10th August 2012[1].

The use of pesticides in agriculture has been a mixed blessing. Pesticides as a plant protection tool helped in increased food production, but on the other side of the success, the problems such as presence of pesticide residues in food and feed, environmental pollution, pest resistance, pest resurgence, outbreak of secondary pests, killing of non-targets including natural enemies, pollinators etc were also noticed, and out of all these, the presence of pesticide residues is a cause of serious concern by the general public. These food safety issues are major concern for all consumers, and hence all National Governments are focusing on the monitoring of pesticide residues in various foods, and excess of regulatory limits. The persistence of pesticides and their residues in commodity at harvest/foods depends on several factors such as nature and amount of pesticide used, number of applications, type of crop, method of application, weather condition, interval between application and harvest etc. In addition, the residues can also occur as a result of circumstances not designed to protect the crop, and soil containing residues of persistent pesticides.

The pesticides residues in food are rigorously regulated in many countries to address the domestic food safety and trade issues/disputes under SPS Agreement of WTO. At International level, the Codex Committee on Pesticide Residues (CCPR) is responsible for establishing maximum limits for pesticide residues in specific food items or in groups of food; establishing maximum limits for pesticide residues in certain animal feeding stuffs moving in international trade where this is justified for reasons of protection of human health; preparing priority lists of pesticides for evaluation by the Joint FAO/WHO Meeting on Pesticide Residues (JMPR);

considering methods of sampling and analysis for the determination of pesticide residues in food and feed; considering other matters in relation to the safety of food and feed containing pesticide residues and; establishing maximum limits for environmental and industrial contaminants showing chemical or other similarity to pesticides in specific food items or groups of food. In India, pesticide residues in foods are regulated under Prevention of Food Adulteration Act, 1954 and now under Food Safety and Standards Act, 2005, and Insecticide Act, 1968.

In order to exploit full potential of pesticides in agriculture and public health programme without adversely affecting the environment, it is essential to study the facts about pesticide behavior, their persistence/dissipation under tropical Indian conditions. Pesticide residues in food commodities and their entry into the food-chain has become a major cause of concern all-over the world, to ensure the safety to the consumer. In view of the above, Government of India through Indian Council Agricultural Research started an All Indian Coordinated Research Project on Pesticide Residue, re-designated as All India Network Project on Pesticide Residues, during the year 1984-85, with a mandate to develop protocols for the safe use of pesticides by recommending 'good agricultural practices', based on multi-location 'supervised field trials' and "Fixing MRLs" and the pre-harvest interval (PHI)/safe waiting period through the analysis of residue data collected at various intervals. This project involves regular monitoring of pesticide residues in biotic and abiotic component of the environment to know the extent of contamination by the indiscriminate and non-judicious use of pesticides so that the Government could initiate necessary remedial actions. There are 12 coordinating centres of which ten are located in State Agricultural Universities (PAU, Ludhiana; RAU, Jaipur; BCKVV, Mohanpur; ANGRAU, Hyderabad; CCSHAU, Hisar; AAU, Anand; Dr YSPUHF, Solan; CSAUAT, Kanpur; KAU, Vellayani; MPKV, Rahuri), one in Central Agricultural University (Mizoram) and two in ICAR Institutes (IARI, New Delhi; IIHR, Bangalore). The Project Coordinating Cell is located in the Division of Agricultural Chemicals, Indian Agricultural Research Institute, New Delhi.

After the formation of WTO (World Trade Organization), presence of the residues above the permissible level is also a major bottleneck in the acceptance of food commodities by the importing counties under WTO agreement on the Application of Sanitary and Phytosanitary Measures (SPS Agreement), and to overcome the trade barriers at International level with respect to Food Safety issues, it is important to know the status of pesticide residues so as to produce the scientific data in dealing such issues, as well as protection of health of the consumers through food safety. Later, on the recommendation of Joint Parliamentary Committee (JPC) on Pesticide Residues during 2003, the Department of Agriculture and Cooperation (DAC), Ministry of Agriculture has started the central sector scheme, "Monitoring of Pesticide Residues at National Level" during 2005-06 with the participation of 21 participating laboratories representing six ministries under supervision of The Network Coordinator, AINP on Pesticide Residues as the Member Secretary.

Through various projects and schemes, Government of India, monitors the various food commodities for food safety, with special reference to pesticide residues through Indian Council of Agricultural Research, and the data generated is communicated to various line departments for sensitizing the safe and judicious use of pesticides in Agriculture, so as to limit the pesticide residue contamination in foods, for consumer safety.

Pesticide Use and Regulations

More than 2.6 million tonnes of active ingredients of pesticides are used worldwide. Roughly 85 per cent of this consumption is used in agriculture. About three-quarters of pesticide are used in developed countries, mostly in Europe and Japan. India's consumption of pesticide is only 2 per cent of the total world consumption. During the last five decades the consumption of pesticides in India has increased several hundred folds till 1991-92 (Figure 9.1) from 2353 metric tonnes (MT) in 1955-56 to 75,000 MT in 1990-91[2]. However, their consumption steadily declined thereafter to 41,820 MT during 2009-10, primarily due to: i) banning of high dose persistent organochlorine pesticides such as DDT, aldrin, heptachlor etc and ii) introduction of new generation pesticides such as neo-nicotinoid, that are effective at doses as low as 20 g a.i./ha. The per hectare consumption of pesticide in India is 381 g which is low as compared to the world average of 500 g. Low consumption in India can be attributed to fragment land holdings, low level of

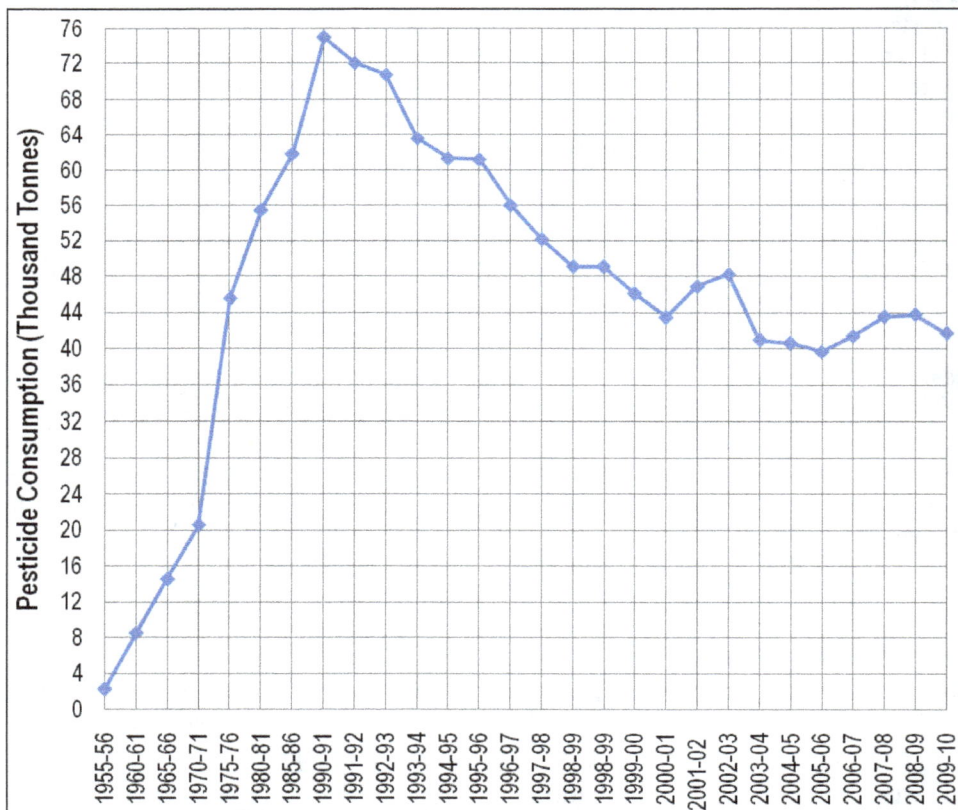

Figure 9.1: Pesticide Consumption (Technical Grade) in India during 1955-2010.

irrigation, dependence on monsoons, low awareness among farmers about the benefits of usage of pesticides etc. India, being a tropical country, the consumption pattern is also more skewed towards insecticides, which accounted for 64 per cent of the total pesticide consumption in 2006-07. Fruits and vegetables consume the highest amount of pesticides (26 per cent) in the world, followed by cereals (15 per cent), maize (12 per cent), rice (10 per cent) and cotton (8.6 per cent). In India however 45 per cent of the total pesticide consumption is on cotton crop followed by rice (22 per cent), vegetables (9 per cent) and pulses (4 per cent) and the trend is now changed after the introduction of transgenic cottons.

The area under plant protection is continuously increasing, but still only 25-30 per cent of total cultivated area is under pesticide cover. The approach to control pests and increase the yield of crops in developed countries is by controlling weeds that compete with crop for nutrients and harbor insect pests and diseases. Therefore out of the total pesticide used, 36 per cent are herbicides, whereas insecticides are 25 per cent and fungicides 10 per cent. In developing countries, however, insecticides constitute a major portion of pesticide consumption but only 12 per cent herbicides are used.

Before a pesticide is sold in the market, it is to be registered under the Insecticides Act, 1968 (An Act to regulate the import, manufacture, sale, transport, distribution and use of insecticides with a view to prevent risk to human beings or animals, and for matters connected therewith). Central Insecticide Board and Registration Committee (CIB and RC), facilitate the registration of safe, efficacious and quality pesticides for domestic use and export, and also disseminate information to State Governments and other concerned departments/agencies for effective implementation of Insecticide Act and Rules framed there under. Besides Insecticide Act, 1968 and Insecticide Rules, 1971, the pesticides/pesticide residues in foods are regulated under various acts, specially under Prevention of Food Adulteration Act (PFA), 1954 and Rules 1955, but after the establishment of Food Safety and Standards Authority of India (FSSAI) under the Food Safety and Standards Act, 2006 as a statutory body for laying down science based standards for articles of food and regulating manufacturing, processing, distribution, sale and import of food to ensure safe and wholesome food for human consumption, all the issues related to food safety *vis-a-vis* pesticide residues, and Maximum Residue Limits (MRLs) are dealt only under Food Safety and Standards Act, 2006 and Food safety and standards (contaminants, toxins and residues) regulation, 2011[3].

Pesticide Dissipation/Residues and Regulatory Guidelines

Pesticide under registration should be tested for bio-efficacy, toxicity and residues as per Insecticide Act, and with respect to studies on pesticide dissipation/residues in/on crops, the data generated by All India Network Project on Pesticide Residues, Indian Council of Agricultural Research, help Government of India, to assess the persistence of pesticide in/on crops, soil, and also to recommend the Maximum Residue Limit, and Pre-Harvest Interval (PHI) for food safety. Since the inception of the project, more than 450 multi-location supervised field trials have been conducted for the safety evaluation of pesticides after the application on field crops following Good Agricultural Practices (GAPs). Various MRL's have been fixed and expansion of label claim has been notified by the Central Insecticide Board and Registration Committee (CIB and RC) on the data generated under multi-location supervised field trials[4]. Recently, the label claim extension of imidacloprid 70 WG formulation on cotton and rice; imidacloprid 200 SL on pea; imidacloprid 70 WG on okra; spiromesifen 480SC on tea, brinjal, chilli, apple; flubendiamide 480 SC on cabbage, tomato; flubendiamide 240 + thiacloprid 480 (SC) on tomato; thiacloprid 240SC on apple, tea; trifloxystrobin + tebuconazole 75 WG on grape, apple, fipronil 80WG on chilli have been approved by the CIB and RC for their commercial use in the country. The multilocation supervised field trials have been conducted for the combination formulation of fluopicolide 6.25 per cent + propamocarb hydrochloride 62.5 per cent - 68.75 SC in/on tomato; quizalofop-p-tefuryl (post emergence herbicide) on cotton; combination product quizalofop ethyl 10 per cent EC + chlorimuron ethyl 25 per cent WP on soybean; flubendiamide 480 SC on brinjal and cardamom; thiacloprid 240 SC on brinjal; fenoxaprop-p-ethyl 9 EC on onion; combination formulation fenamidone 50 per cent + mancozeb 10 per cent - 60 WG on gherkin; combination formulation spirotetramat 120+ imidacloprid120- 240 SC on brinjal, okra and mango.

To address the food safety issues, under FSSAI, the data generated by AINP on Pesticide Residues, ICAR during last few years, has helped the Ministry of Health and Family Welfare, Government of India, for fixing the MRLs. Table 9.1 elucidates the achievements in this regard.

Table 9.1. Maximum Residue Limits (MRLs) Fixed Under PFA/FSSAI Based on the Data Generated by AINP on Pesticide Residues, ICAR, during last Few Years

Pesticide	Crop	MRL Fixed Under PFA/FSSAI (mg/kg)
Flubendiamide 480 SC	Brinjal	0.10
Spiromesifen 240 SC	Chilli	0.50
Spiromesifen 240 SC	Brinjal	0.50
Spiromesifen 240 SC	Apple	0.01
β-cyfluthrin + Imidacloprid	Okra	
Imidacloprid		2.00
β-cyfluthrin		0.01
β-cyfluthrin + Imidacloprid	Brinjal	
Imidacloprid		0.01
β-cyfluthrin		0.01
Thiacloprid 240 EC	Tea	5.00
Flubendiamide 480 SC	Cabbage	0.05
Flubendiamide 480 SC	Tomato	0.07
Spiromesifen 240 SC	Tea	1.00
Imidacloprid 70 WG	Okra	2.00

Contd...

Table 9.1–*Contd...*

Pesticide	Crop	MRL Fixed Under PFA/FSSAI (mg/kg)
Flubendiamide 480 SC	Chickpea	0.10
Spiromesifen 240 SC	Okra	0.10
Imidacloprid 200 SL	Grapes	1.00
Imidacloprid 350 SC and 70 WG	Paddy	0.05
Spiromesifen 240 SC	Tea	1.00
Thiacloprid 240 SC	Apple	0.05
Imidacloprid 200 SL	Chilli	0.30
Imidacloprid 70 WG	Cucumber	0.20
Imidacloprid 70 WG	Cotton	0.05
Deltamethrin 100 EC	Chilli	0.50
Fipronil 80 WG	Grapes	0.01
Flubendiamide 480 SC	Chilli	0.02
Fipronil	Chilli	0.001
Spiromesifen 240 SC	Tomato	0.30
Fipronil 80 WG	Cabbage	0.001
Tebuconazole 60 FS	Wheat	0.07
Thiacloprid 240 SC	Tea	5.00
Thiacloprid 240 SC	Brinjal	0.30
Deltamethrin 10 EC	Tea	2.00
Quizalofop ethyl 5 EC	Black gram	0.01
Cartap hydrochloride	Rice	0.50
Quizalofop ethyl 5 per cent EC	Onion	0.01
Quizalofop ethyl 5 per cent EC	Soybean	0.05
Quizalofop ethyl 5 per cent EC	Groundnut	0.10
Quizalofop ethyl 5 per cent EC	Cotton	0.10
Flusilazole 12.5 per cent + Carbendazim 25 per cent SE	Rice	
Flusilazole		0.01
Carbendazim		0.50

The project has conducted 160 multi-location supervised field trials and generated pesticide residue data of the deemed registered pesticides for fixation of MRL and expansion of label claim. Under this program, residue data has been generated for quinalphos on cauliflower, onion, citrus fruit, brinjal, maize, mustard and soybean; methyl parathion in black bram, soybean, mustard, cotton, paddy; oxydemeton methyl in citrus fruit, chillies, mustard, cotton; carbaryl on grapes, sesamum, kinnow; dichlorvos on mustard and dimethoate on tea.

Pesticide Residues in Fruits and Vegetables: A National Initiative and Study Results

Pesticide Residues in/on food is a matter of concern under food safety, and Government of India, regulate the pesticide residue contamination in various food items through Prevention of Food Adulteration Act, and now through Food Safety and Standards Act, 2005. Various Organizations in India; Indian Council of Agricultural Research (ICAR) institutions; State Agricultural Universities; Central Insecticide Laboratory (CIL); Indian Institute of Grain Storage; Indian Council of Medical Research (ICMR) Institutions; and Council of Scientific and Industrial Research (CSIR) and Bhabha Atomic Research Centre, and other research groups have engaged in monitoring of pesticide residues in food commodities and environmental samples in individual capacity primarily for academic purposes. Such studies were overlapping in many areas and differed from one another in their results. Due to increasing public awareness and legalities involved in pesticide residues in food commodities there was a need to harmonize the monitoring of pesticide residues in the country. Further, after implementation of SPS Agreement under WTO, on the recommendation of Joint Parliamentary Committee (JPC) on Pesticide Residues during 2003, the Department of Agriculture and Cooperation (DAC), Ministry of Agriculture has initiated the central sector scheme, "Monitoring of Pesticide Residues at National Level" during 2005-06 involving Ministry of Agriculture, Ministry of Health, Ministry of Chemicals and Fertilizers, Ministry of Commerce, Ministry of Environment and Forest and State Agricultural Universities, and The Project Coordinating Cell, AINP on Pesticide Residues, IARI, New Delhi of ICAR as Nodal Centre.

These data help focus governmental interventions such as Integrated Pest Management and other suitable measures in regions of high prevalence of pesticide residues. The prime objectives of research and regulatory programs of the country are the achievements of pest control without injury to man, animals, plants, soil and other values in man's total environment. This program helps the authorities at various levels to sensitize the stake holders on pesticide use/safe use, pesticide residue and food safety, and regulatory issues related to pesticide residues.

The food commodities like milk, cereals, vegetables, fruits, meat and marine products, fish samples and water samples in different agricultural regions of the country with special emphasis on items of daily consumption as well as export were analyzed at weekly/monthly intervals, drawing farm gate and market samples of each commodity. Uniform methodology is being followed by all the participating laboratories for sampling, analysis of residues and reporting of the results. The results are being confirmed with the help of Gas Chromatograph-Mass Spectrometry (GC-MS) or Liquid Chromatograph-Mass Spectrometry (LC-MS).

During October, 2006 to March, 2012 a total of 23,129 samples of the vegetables (tomato, cabbage, brinjal, capsicum, okra and cauliflower) and 8859 samples of fruits (apple, banana, grape, orange, pomegranate, guava and mango) were collected and analyzed for the possible presence of pesticide residues across various parts of the country under the scheme. Pesticide residues were detected in 3547 (15.2 per cent) samples of vegetables and 779 (8.5 per cent) samples of fruits, out of which 696 (3 per cent) samples of vegetables and 74 (0.9 per cent) samples of fruits were found to contain residues above their Prevention of Food Adulteration (PFA)/Codex notified Maximum Residue Limit (MRL). The data on pesticide residues contamination in vegetables and fruits are presented in Tables 9.2 and 9.3 and Figures 9.2 and 9.3, respectively[5].

Table 9.2: Monitoring Data on Pesticide Residue Contamination in Vegetables Collected from Different Agro-Climatic Zones of India and States of India during 2006-2012

Year	Number of Samples Analyzed	Number of Samples with Detected Residues	Number of Samples Detected Residues above MRL
Oct, 2006-March, 2007	761	299	61
April, 2007-March, 2008	2529	699	113
April, 2008-March, 2009	3989	531	117
April, 2009-March, 2010	4239	417	84
April, 2010-March, 2011	5170	593	113
April, 2011-March, 2012	6441	1008	208
Total	**23,129**	**3,547 (15.2 per cent)**	**696 (3 per cent)**

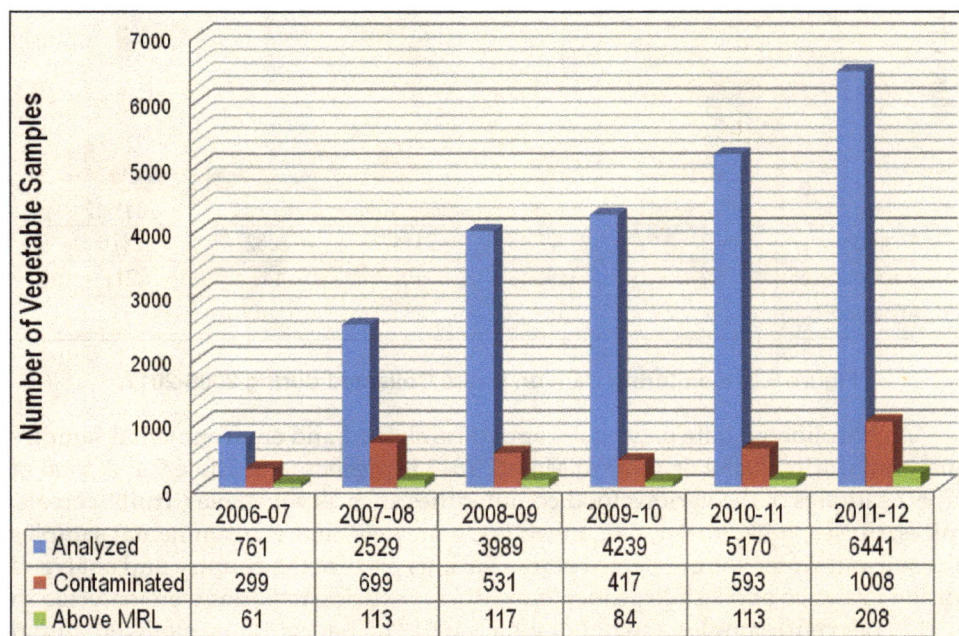

	2006-07	2007-08	2008-09	2009-10	2010-11	2011-12
■ Analyzed	761	2529	3989	4239	5170	6441
■ Contaminated	299	699	531	417	593	1008
■ Above MRL	61	113	117	84	113	208

Figure 9.2: Monitoring Data on Vegetables Collected during 2006-2012.

Table 9.3: Monitoring Data on Pesticide Residue Contamination in Fruits Collected from Different Agro-Climatic Zones of India and States of India during 2007-2012

Year	Number of Samples Analyzed	Number of Samples with Detected Residues	Number of Samples Detected Residues above MRL
April, 2007-March, 2008	471	77	4
April, 2008-March, 2009	2042	159	20
April, 2009-March, 2010	2114	187	18
April, 2010-March, 2011	2062	145	20
April, 2011-March, 2012	2170	211	12
Total	**8,859**	**779 (8.7 per cent)**	**74 (0.9 per cent)**

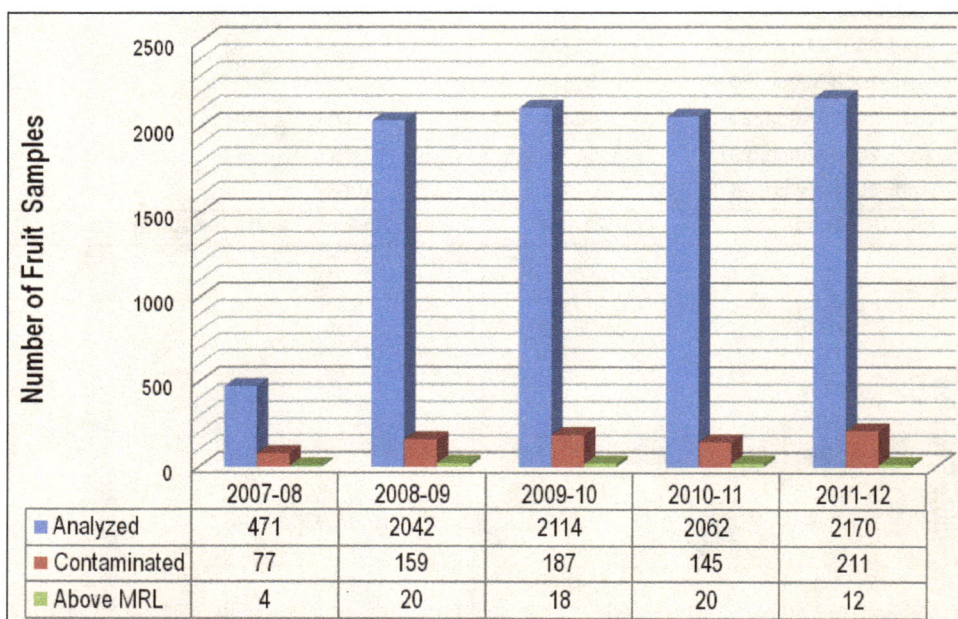

	2007-08	2008-09	2009-10	2010-11	2011-12
■ Analyzed	471	2042	2114	2062	2170
■ Contaminated	77	159	187	145	211
■ Above MRL	4	20	18	20	12

Figure 9.3: Monitoring Data on Fruits Collected during 2006-2012.

A consolidated data on various agricultural, food and environmental samples analyzed during October, 2006 to March, 2012 is presented in Table 9.4. A total of 67,992 samples of the various food commodities such as vegetables, fruits, cereals, spices, pulses, milk, butter, fish, meat, tea, honey etc. and environmental samples like soil and water were collected across various parts of the country and analyzed for the presence of possible pesticide residues. Pesticide residues were detected in 6281 (9 per cent) samples and out of which 1051 (1.5 per cent) samples were found to contain residues above their Prevention of Food Adulteration (PFA)/Codex notified Maximum Residue Limit (MRL).

Table 9.4: Monitoring Data on Pesticide Residue Contamination in Agricultural Commodities, Food and Environmental Samples Collected from Different Agro-Climatic Zones of India and States of India during 2006-2012

Year	Number of Samples Analyzed	Number of Samples with Detected Residues	Number of Samples Detected Residues above MRL
Oct, 2006-March, 2007	1061	347 (32.7 per cent)	61 (5.75 per cent)
April, 2007-March, 2008	7089	1279 (18.0 per cent)	152 (2.1 per cent)
April, 2008-March, 2009	13348	1039 (7.8 per cent)	212 (1.6 per cent)
April, 2009-March, 2010	14225	899 (6.3 per cent)	155 (1.1 per cent)
April, 2010-March, 2011	15321	1049 (6.8 per cent)	181 (1.2 per cent)
April, 2011-March, 2012	16,948	1668 (9.8 per cent)	290 (1.7 per cent)
Total	**67,992**	**6,281 (9 per cent)**	**1051 (1.5 per cent)**

Concluding Remarks

National supervised field studies on Pesticide dissipation/residues in various Crops as per Good Agricultural Practices conducted by All India Coordinated Research Project of Indian Council of Agricultural Research help to fix Maximum Residue Limits as per Insecticide Act, 1968 and Food Safety and Standards Act, 1954. The MRLs help India to address the issues of Food Safety both at domestic and International level, so as to regulate pesticide residues in food commodities and also eliminate the trade barriers, following Pre-Harvest Intervals. Pesticide residues in foods are highly regulated by FSSAI in India, and the National monitoring of pesticide residue contamination in fruits and vegetables, conducted during 2006-2012, indicate that around 15 per cent of different vegetables are detected with one/more pesticide residues, and the matter of concern is that about 3 per cent vegetables were detected pesticide residues above legal limits *i.e.* MRL. However, in case of fruits, out of 9 per cent pesticide residue detected samples, about 0.9 per cent samples are contaminated with pesticide residues above MRLs. As the food consumption habits are changing due to globalization, for example, the raw vegetable consumption is increasing (salads), the risk of pesticide consumption is to be considered and pesticide residues are to be regulated through education of all stake holders on safe and proper use of pesticides, and pesticide residues implications on food safety and human health.

References

Annual Reports (2005-2011) of All India Network Project on Pesticide Residues, Indian Council of Agricultural Research.

Annual Reports (2006-2011) of Central Sector Scheme "Monitoring of Pesticide Residues at National Level", Department of Agriculture and Cooperation, Ministry of Agriculture, Government of India.

Food safety and standards (contaminants, toxins and residues) regulation, 2011 under Food Safety and Standards Act, 2006 (http://www.fssai.gov.in).

Insecticides/pesticides registered under section 9 (3) of The Insecticide Act, 1968 for use in the country (http://cibrc.nic.in).

Production and use of Agricultural Inputs in India, Directorate of Economics and Statistics, Department of Agriculture and Cooperation, Ministry of Agriculture, Government of India (http://agricoop.nic.in/Agristatistics.htm).

Chapter 10

Pest Forecasting and Modelling

Y G Prasad

ICAR-Central Research Institute for Dryland Agriculture,
Hyderabad – 500 059, Telangana State, India
E-mail: ygprasad@gmail.com

ABSTRACT

Integrated pest management (IPM) is a system that emphasizes appropriate decision making, information intensive and depends heavily on accurate and timely information for field implementation by practitioners. Pest forecasts are an important component of the broad IPM philosophy. Pest forewarnings provide lead time for impending attacks and thus minimise crop loss and optimize pest control leading to reduced cost of cultivation. Pest populations fluctuate in timing and intensity of their occurrence under the influence of several macro and micro-climatic factors depending on the location and season. Pest prediction models are useful ways of synthesizing biological or empirical data as influenced by the environment. Data needed for building and validating models are collected through a variety of monitoring tools and field scouting of population level and crop damage assessment. In this paper, empirical and process-based modeling approaches adopted by researchers to achieve the common aim of providing pest forecasts are discussed. In recent years, generic simulation modeling tools have become available for model building and running. Farmers are mainly interested in current pest severity data, preferably for their localities to aid their decision-making in crop protection. Operational pest forecasting through networks for near real time collection of weather data for use in estimation of pest development based on forecast models and computer-based decision support systems is the way forward.

Introduction

Integrated pest management (IPM) is a system that emphasizes appropriate decision making, knowledge intensive and depends heavily on accurate and timely information for field implementation by practitioners. Pest forecasts are an important component of the broad IPM philosophy. Pest and disease outbreaks sometimes result in total crop failure and adversely affect farmers income. Pest forewarnings provide lead time for impending attacks and thus minimise crop loss and optimize pest control leading to reduced costs of cultivation.

Insects and crops have coevolved over a long period of time. As a result insects have adapted themselves to a particular crop or group of crop plants and also have specialized feeding niches. Pest forecasts will have to take several factors into consideration while being devised in the form of model outputs and disseminated as alerts to farmers. Many pests and diseases are regular in their appearance at susceptible stages of crop growth *e.g.*, pod borer, *Helicoverpa armigera* (Hubner) on pigeonpea at flowering period. The pest appears year after year with its population fluctuating around a seasonal mean. Some pests and diseases are endemic to certain locations (*e.g.*, rice gall midge) while some are cyclical or sporadic in their appearance (e.g, Bihar hairy caterpillar, *Spilosoma obliqua* Walker on sunflower). Some are characterized as outbreak pests or diseases because of their potential to rapidly increase in their abundance in a short span triggered by certain congenial factors leading to widespread epidemics *viz.*, brown planthopper, *Nilaparvatha lugens* Stal and blast in rice; *Spodoptera litura* (Fab.) on soybean and cotton; mealybug, *Paracoccus marginatus* Williams and Granara de Willink on papaya and several other crops. Species which have a wider host range have better chances of carry over to the next season or year. Species which are monophagous or oligophagous have resting or dormant phases in their lifecycles to tide over unfavourable weather condtions or lack of a susceptible host or crop stage. Pest and disease appearance on a crop follows its life cycle and is also intricately connected to the crop phenology. Some insects complete only a single generation a year (*e.g.* red hairy caterpillar, *Amsacta albistriga* Walker) while others have multiple and often overlapping generations. Similarly diseases could be monocyclic or polycyclic in their epidemiology and generally require a latent period to cause infection *e.g.*, late leaf spot disease on groundnut. Pest and disease forecast models need to consider these intrinsic attributes of the insect pest or disease and also determining enviromental and host factors.

Factors Influencing Pest Abundance

Pest populations fluctuate in timing and intensity of their occurrence depending on location and season. Mostly they tend to fluctuate over a mean level. This average population over time when computed across several years results from the sum of action of all positive and negative factors influencing pest populations. Pests of host plants in undisturbed habitats such as forestry have their natural cycles in response to their ecosystem interactions and are most likely to attain equilibrium points in their population levels (Figure 10.1). Pests of agro-ecosystems fluctuate within and between short spans of time *i.e.*, season and off-season, as they experience rapidly changing environments due to changes in cropping systems and a host of management interventions (Figure 10.2). As a result crop pests show greater degree of instability in population levels. An abnormal increase in the numbers of a species on any crop in a short span of time over a large area leads to a significant impact on production and productivity and is considered as a pest outbreak or epidemic. Many factors are responsible for such an outbreak. Foremost among them is the prevalence of congenial weather conditions for insect multiplication and rapid buildup. Growing susceptible cultivars of crop plants in monocultures on a large

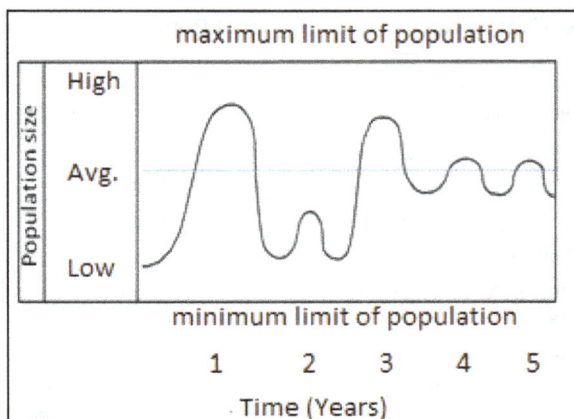

Figure 10.1: General Equilibrium Position of an Insect in an Undisturbed Ecosystem.

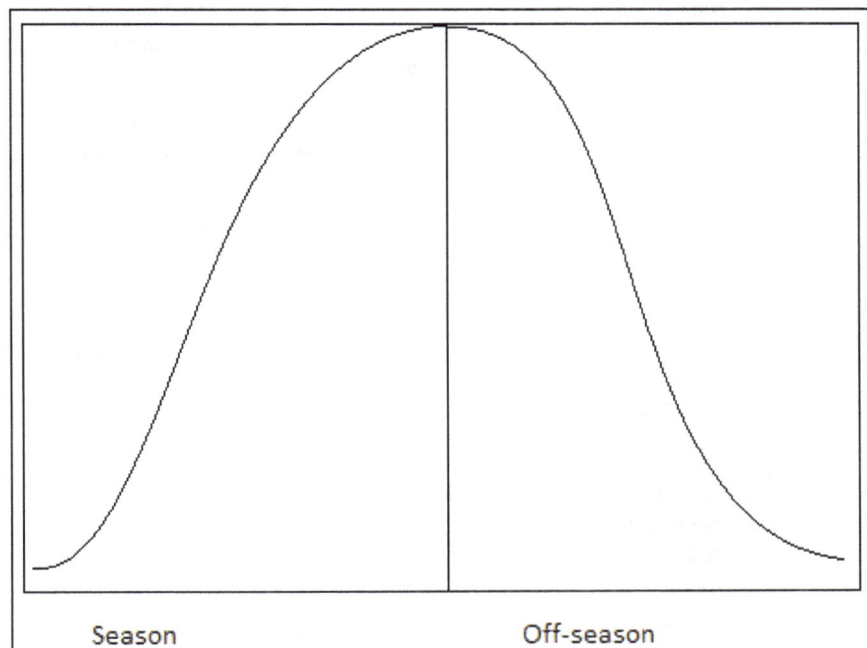

Figure 10.2: Population Fluctuation in Agroecosystems.

scale year after year may help insect pests to multiply and spread fast. Sometimes indiscriminate use of insecticides for pest control disrupts the ecological balance between phytophagous insects and their natural enemies leading to pest outbreaks apart from causing pest resurgence and pesticide resistance. Invasive pest species exhibit outbreak tendency due to lack of natural enemies, which are crucial in regulation of pest populations. In general, insects are influenced by their intrinsic attributes and extrinsic factors such as weather and crop factors (Table 10.1).

Table 10.1: Factors Influencing Pest Population Dynamics

Factors	Parameter	Influence on
Pest attributes	Life stages and their duration	Generation time and number
	Conditions for stage transfer	Diapause
	Timing of reproduction	Pest onset
	Fecundity	Pest abundance
	Survival of stages	Pest numbers
Weather	Temperature (maximum and minimum)	Insect developmental rate, survival, fecundity, diapause
	Humidity (maximum and minimum)	Survival, egg hatch
	Rainfall (amount and distribution)	Adult emergence, oviposition, spread, mortality (wash-out)
	Bright sunshine hours/Photoperiod	Food quality, oviposition, diapause
	Wind speed and direction	Dispersal and migration
	Microclimatic parameters (canopy temperature, humidity, leaf wetness, soil temperature)	Development and rate of spread, mortality Plant diseases: spore germination, infection and sporulation
Crop	Genotype	Rate of development, survival
	Phenology	Pest appearance and intensity
	Crop diversity, geometry and stand	Insect diversity, abundance, survival and dispersal
	Nutrition	Fecundity
	Time of sowing	Pest onset
	Cropping system practices	Off-season survival, carry over
	Density dependent mortality factors	Pest regulation

On a global scale, seasonal temperatures and rainfall patterns constitute major factors that determine the distributions of organisms in space (Birch, 1957). Tropical insects on the average have the same annual variability as insects from temperate zones, but insect populations from dry areas, temperate or tropical, tend to fluctuate more than those from wet areas (Wolda, 1978).

In nature pests are regulated by their natural enemies: parasitoids, predators and pathogens which are again influenced by bio-physical factors (Hence *et al.*, 2007; Thomson *et al.*, 2010,). Therefore, precise understanding of population dynamics can result from comprehensive ecological studies. However, despite best efforts gaps in pest ecological databases exist due to the complexity of interactions among the ecosystem components.

Pest Monitoring Tools

Monitoring for pests is a fundamental first step in designing a proper integrated pest management (IPM) program. Pest monitoring data help build and validate forecast models. Pests are monitored through a variety of monitoring tools such as pheromone traps, light traps, coloured sticky traps, pitfall traps and suction traps. Among the monitoring tools, the most popular and widely used are sex pheromone traps for selective monitoring of individual flying species and light traps for flying species that are attracted to light. While adult males are mostly caught in sex pheromone traps, adults of both sexes are trapped in light traps. The timing of adult male catches in the trap indicates the start of the pest flight activity in the area. This information is important for some pests as it is used as bio-fix date for accumulation of heat units above a base temperature in phenology models or sustained first flight in others (Knutson *et al.*, 2010). Trap capture data serve several purposes (Table 10.2). Linear relationship between male catches in sex pheromone traps and growing degree-days is possible after appropriate transformation of variables (Gallardo *et al.*, 2009) and in some cases variability is better explained by including other variables related to density of host plants or suitable plant parts (Jennifer and Allen, 2007). Peak pheromone trap captures are often correlated with associated weather to identify positive or negative influence of weather parameters on moth activity and pest build-up (Gwadi *et al.*, 2006, Reardon *et al.*, 2006; Monobrullah *et al.*, 2007, Prasad *et al.*, 2008). However, trap catches and weather may not necessarily serve as predictors of future abundance of certain species in cropping regions (Baker *et al.*, 2010). Light traps have been widely used for monitoring population dynamics of Lepidoptera and Coleoptera (Wolda, 1992; Watt and Woiwod, 1999; Kato *et al.*, 2000). When compared to other sampling methods, light trap sampling was more efficient for Lepidoptera population dynamics (Raimondo *et al.*, 2004). However, many factors affect catches of insects in light traps (Bowden, 1982).

Modelling Insect Populations

Pest prediction models are mathematical descriptions of biological or empirical data as influenced by the environment. They summarize biophysical relationships in the form of equations and provide a basis to test the significance of this relationship. Important points for consideration in any model development are: the level of detail at which a given model is to be developed as the level of detail is linked to the objective and data availability to develop and run the model. Models can range from strictly empirical to most complex and sophisticated descriptive models. A model may be discrete or continuous, static or dynamic, and deterministic or stochastic.

Building and running pest forecast models require access to weather and climate data, in addition to pest and plant data. Models usually require as inputs, measurements of temperature, rainfall and humidity, although other variables may be required either as direct inputs or in computing values for variables not measured. Weather variables need to be measured at the field level, at regional stations, or on a broader scale depending on the need. For many farm management actions, data representative of the field conditions are expected and hence data is taken from automatic weather stations or the nearest observatory.

Various types of pest/disease incidence data (trap catches, population counts and crop damage assessments) could be made use of. Long-term data are preferable as it better captures the patterns in relationships. However, historical data collected from different sources suffer from several inadequacies: lack of ancillary data such

Table 10.2: Monitoring of insect Pests and Use in Forecasts

Trap	Purpose	Reference
Sex pheromone traps	Bio-fix or starting dates in insect phenology models for accumulating degree-days for prediction of egg hatch and first larval damage; testing and field validation of models	Jennifer and Allen, 2007; Gallardo *et al.*, 2009; Knutson *et al.*, 2010,
	Adult population monitoring, Initiating field scouting for immature stages, Timing of insecticide treatments	Wall *et al.*, 1987; Guerrero and Reddy, 2001, Witzgall *et al.*, 2010
	Detection and tracking of low density populations of invasive insect pests	El-Sayed *et al.*, 2006; Liebhold and Tobin 2008
	Determination of outbreak area and effectiveness of eradication campaigns	Cannon *et al.*, 2004
	Population monitoring at regional level via a network of sites for early detection and area-wide IPM	Ayalew *et al.*, 2008; Hopkinson and Soroka, 2010
Light traps	Prediction of moth emergence, adult activity and assessment of population fluctuations over time	Lewis, 1980; Zou *et al.*, 2004; Anonymus, 2009
	Prediction of population size based on catches	Odiyo *et al.*, 1979; Raimondo *et al.*, 2004; Zalucki and Furlong, 2005
	Monitoring seasonal migration of *Helicoverpa armigera* Hubner and migration waves of *Nilaparvata lugens* Stal	Zhu *et al.*, 2000; Feng *et al.*, 2009
	Operational pest forecasting	Drake *et al.*, 2002
	Validation of simulation models	Regi and Chander, 2008
Suction traps	Migration monitoring of aphids and correlation with weather data	Klueken *et al.*, 2009

as sowing times, crop damage assessments, and lack of time-series data, gaps in data and lack of uniformity in data format across years.

There are several approaches adopted by modeling groups to achieve the common aim of providing pest forecasts (Table 10.3). These approaches broadly fall into two categories: empirical and process-based. Empirical approaches involve estimating pest and disease incidence and intensity through experimentation and surveys on crops not subjected to control interventions and establishing relationships with concurrent, prevailing weather and/or past weather factors. These studies could be conducted at single stations in which the emphasis is on delineation of differences in meteorological conditions in epidemic and non-epidemic years or multi-station studies in which the emphasis is on delineation of meteorological conditions leading to changes in periods and intensity of infestations. A multi-station study is preferred as it facilitates corroboration of the general surmises and leads to maximization of data in a short period if observations are recorded on crop stands sown at periodic intervals at a number of stations (Venkataraman and Krishnan, 1992). It should be noted that findings from empirical field studies can straight away be applied in climatologically analogous areas but can give misleading results when applied to other areas.

Table 10.3: Empirical and Process-Based Approaches to Pest Modelling

Parameter	Statistical Approach	Process-based Approach
Driving variables	Weather	Rate and state variables, bio-ecological and physical processes
Data source	Surveys, experiments	Experiments based on prior assumptions
Analytical methods	Correlation and regression, statistical, data driven, quantitative	Descriptive, mathematical, data and knowledge driven, mechanistic, deterministic or stochastic, iterative
Model building sophisticated	Data intensive, straight forward	Knowledge intensive, complex and
Model validation	Time consuming	Time saving simulations
Application domain	Location specific, applicable to matching climates and ecologies	Universal, applicable across environments
Operational use	Model elements synthesized into computer programs or online tools	Computer based decision support systems

Process-based models are essentially descriptive or mechanistic and are dependent on insect life cycles, ecological and physical processes. These models are complex to build, sophisticated to run and are essentially computer based (Table 10.3). Correlation and regression approach to modeling has been widely adopted in the tropics in Asia. Phenology and processed based modeling is in vogue in temperate and tropical regions of developing countries.

Regression Models

Regression models consider pest/disease incidence variables as dependent and suitable independent variables such as weather variables, crop stages, population of natural enemies/predators etc. is used. The form of the model is:

$$Y = \beta0 + \beta1\ X1 + \beta2X2 + \ldots\ldots\ldots + \beta pXp + e$$

where $\beta0$, $\beta1$, $\beta2$,.......βp are regression coefficients, X1, X2, ..., Xp are independent variables and e is error term. These variables are used in original scale or on a suitable transformed scale such as cos, log, exponential etc.

In case of regression models based on weather indices, using weekly and fortnightly weather variables, suitable indices are worked out which are used as regressors in the model. If information on favourable weather conditions is known, subjective weights based on this information can be used for constructing weather indices. In the absence of such information correlation coefficients between Y and respective weather variable/product of weather initial and final periods for which weather data were included in the model and e is error term. If information on favourable weather conditions is known, subjective weights based on this information can be used for constructing weather indices. In absence of such information correlation coefficients between Y and respective weather variable/ product of weather variables can be used (Agrawal *et al.*, 2004, Desai *et al.*, 2004; Chattopadhyay *et al.*, 2005a, Chattopadhyay *et al.*, 2005b, and Dhar *et al.*, 2007).

In regression analysis, the unfitted errors between a regression model and observed data are generally assumed as observation error that is a random variable having a normal distribution, constant variance, and a zero mean. However, in fuzzy regression analysis, the same unfitted errors are viewed as the fuzziness. Fuzzy regression can be quite useful in estimating the relationship among variables where the availability data are imprecise and fuzzy. Fuzzy regression method is based on minimizing fuzziness as an optimal criterion, which can be achieved by linear programming procedures.

Forewarning models can be developed using the principal component technique as normally relevant weather variables are large in number and are expected to be highly correlated with each other. Using the first few principal components of weather variables as independent variables forecast models can be developed. Discriminant function analysis is applied to time series data on weather variables. For this analysis, a series of data for 25-30 years are required. Based on the pest and diseases variables, data can be divided into different groups – low, medium and high etc. and using weather data in these groups, linear or quadratic discriminant functions can be fitted which can be used to find discriminant scores. Considering these discriminant scores as independent variables and diseases/pest as a dependent variable, regression analysis can be performed. Johnson *et al.* (1996) used discriminant analysis for forecasting potato late blight.

Machine Learning Techniques

Machine learning techniques offer many methodologies like decision tree induction algorithms, genetic algorithms, neural networks, rough sets, fuzzy sets as well as many hybridized strategies for the classification and prediction (Han and Kamber, 2001). Decision tree induction represents a simple and powerful method of classification that generates a tree and a set of rules, representing the model of different classes, from a given dataset. Decision Tree (DT) is a flow chart like tree structure, where each internal node denotes a test on an attribute, each branch represents an outcome of the test and each leaf node represents the class. The top most node in a tree is the root node. For decision tree algorithm and its successor algorithm by Quinlan (1993) are widely used. One of the strengths of decision trees compared to other methods of induction is the ease with which they can be used for numeric as well as non-numeric domains. Another advantage of decision tree is that it can be easily mapped to rules. A decision tree analysis for predicting the occurrence of the pest, *Helicoverpa armigera* and its natural enemies on cotton based on economic threshold level was developed (Pratheepa *et al.*, 2012). The data mining process for construction of decision tree and set of rules is shown in Figure 10.3. The occurrence of cotton bollworm *H. armigera* was greatly influenced by its natural enemies, *viz.*, spiders and *C. carnea* and by abiotic factors. In this analysis, population dynamics of the pest and its natural enemies was studied using Shannon information measure with decision tree induction approach. The developed classification model has the ability to successfully treat 'categorical' variables as well as 'continuous' variables in the database. Pest incidence was classified into two classes, *viz.* low and high based upon economic threshold level (ETL). R^2 for low and high class values was 0.66 and 0.21, respectively. Our studies showed that season influenced the population dynamics of *H. armigera* among all the factors. It was found that the misclassified testing data were 8.8 per cent. The confusion matrix

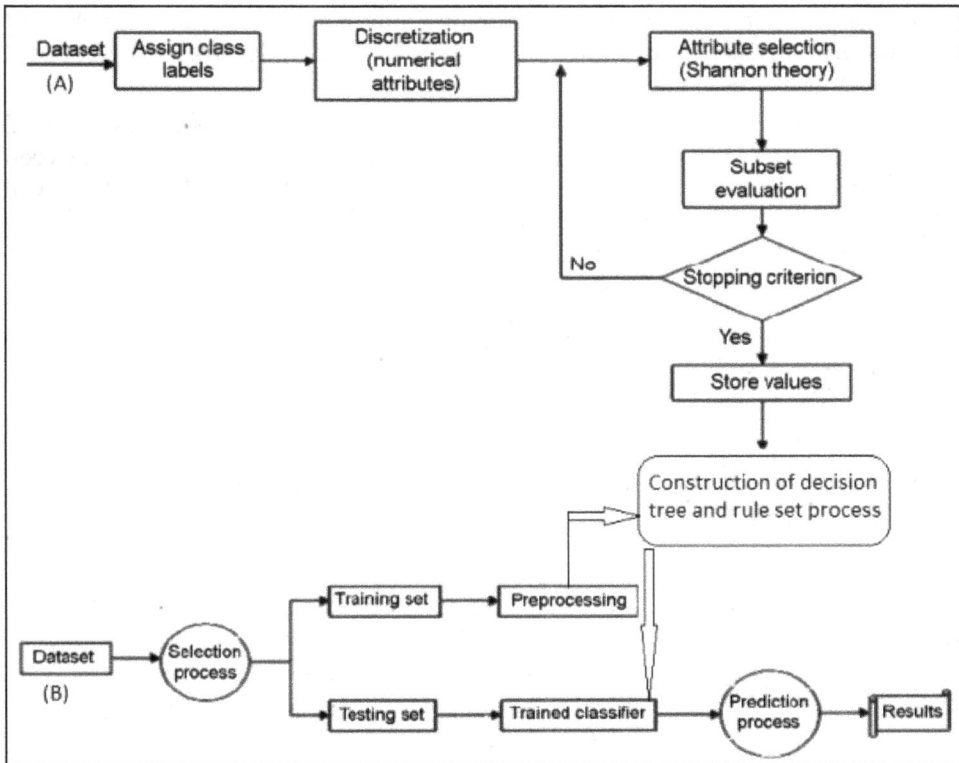

Figure 10.3: Data Mining Process for Construction of Decision Tree (A); Training andTesting Process (B) (Pratheepa *et al.*, 2011).

for the testing set revealed that the classification was done more accurately using the training set. Hence, this approach could be successfully utilized to understand the role of natural enemies and weather factors on the occurrence of *H. armigera* as well as prediction of this pest.

Artificial Neural Networks (ANNs) is another attractive tool under machine learning techniques for forecasting and classification purposes. ANNs are data driven self-adaptive methods in that there are few apriori assumptions about the models for problems under study. These learn from examples and capture subtle functional relationships among the data even if the underlying relationships are unknown or hard to describe. After learning from the available data, ANNs can often correctly infer the unseen part of a population even if data contains noisy information. As forecasting is performed via prediction of future behaviour (unseen part) from examples of past behaviour, it is an ideal application area for ANNs, at least in principle. (Agrawal *et al.*, 2004, Kumar *et al.*, 2010). However, the technique requires a large data base.

Insect Development and Phenology Models

Insects are incapable of their own internal temperature regulation and hence their development depends on the temperature to which they are exposed. The studies of population dynamics often involve modeling growth as a function of temperature. The rate summation methodology has perhaps proven to be the most viable approach to such modeling (Stinner *et al.*, 1974).

The most common development rate model, often called degree-day summation, assumes a linear relationship between development rate and temperature between lower and upper development thresholds (Allen, 1976). This method works well for optimum temperatures (Ikemoto, 2005) (Figure 10.4). The linear model assumes

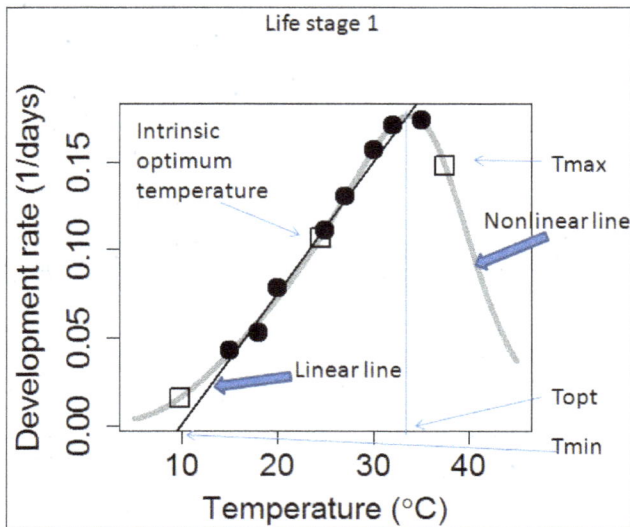

Figure 10.4: Bioclimatic Threshold Estimated from Liner and Nonlinear Fit to Developmental Data at Constant Temperatures.

that rates are proportional to temperature, and since amounts are integrals of rates, the amount of development is the integral of temperature (or a linear function of it) along a time axis and has units of temperature-time (*e.g.*, degree-days). Temperature-dependent development in insects can be approached using either developmental time or rate of development. Rate of development is traditionally utilized because rate models were created from biochemical and biophysical properties (Sharpe and DeMichele, 1977). Most of the earlier models failed to take into consideration variation between individual insects in their rate of development that is responsible for spread of activity of a pest (Regniere, 1984, Phelps *et al.*, 1993). Significant models for modeling the effects of variable temperatures on the development of individual insects within a given population deal with mean rate versus temperature relationships (Wagner *et al.*, 1984a) and distribution of development times (Wagner *et al.*, 1984b, Wagner *et al.*, 1985). Instead of treating rate summation as deterministic quantities, efforts have been made to consider rates as random variables (Stinner *et al.*, 1975). Stochastic approaches to modeling insect development vary in the choice of random variable to be modeled and in the form of the frequency distribution applied to the random variable (Sharpe *et al.*, 1977, Currry *et al.*, 1978). The coefficient of variation of the rate distributions is relatively independent of temperature (Sharpe *et al.*, 1977) indicating that a single temperature-independent distribution of the normalized rate of development adequately describes the distribution at all temperature which has been validated for 80 per cent of 194 sets of published data on 113 species of insects and mites (Shaffer, 1983). Insect species that exhibit seasonality generally have resting phases, diapause or aestivation, in their life cycles which can be accommodated in Monte Carlo simulation modeling (Phelps *et al.*, 1993). Since some temperatures are lethal to organisms, it is obvious that development must be a nonlinear temperature function at the extremes. Non-linear development rate functions based on enzyme kinetics were developed to describe high temperature inhibition (Johnson and Lewin, 1946) and low temperature inhibition (Hultin, 1955) and both the extremes (Sharpe and DeMichele, 1977). Another nonlinear model of temperature-dependent development, (Stinner *et al.*, 1974) utilized a function that is a simple sigmoid curve with an inverted relationship when the temperature is above the optimum. This model, as originally given, assumed symmetry about the optimum temperature, but can be easily modified for asymmetry. A nonlinear model by Logan *et al.* (1976) uses an equation that is asymmetric about the optimum, but becomes negative for very high temperatures. Schoolfield *et al.* (I 981) modified the Sharpe and DeMichele model to enhance its overall utility and simplify parameter estimation. As pointed out by Worner (1992), the interaction of cyclic temperatures with nonlinear development can introduce significant deviations from the linear development rate model, especially in the low and high temperature regions of the development rate function. Stinner's model best fit Russian wheat aphid developmental rate data as judged by mean square error (MSE) and successful convergence when 14 insect developmental models, both deterministic and distributed were tested on laboratory experimental data on development and phenology through simulation (Ma and Bechinski, 2008) using population model design system (Logan and Weber, 1989) software. A new modeling approach to insect reproduction with same shape reproduction distribution and rate

Table 10.4: Bioclimatic Thresholds for Insect Development Estimated by Models Using Stage-Specific Duration Data at Constant Temperatures

Model*	Parameter Estimate	Explanation	Significance
Linear	T_{min} or T_0 or LDT	Lower developmental temperature or base temperature in °C	The temperature at and below which development rate becomes zero
Linear	k or SET	Thermal constant or sum of effective temperatures in degree-days (DD)	Sum of accumulated temperatures above lower developmental threshold (T_{min}) to complete the development of a particular stage or instar; one degree-day is equal to number of degrees above the base temperature in a day
Nonlinear	T_{opt}	Optimum temperature (empirical nonlinear model estimate)	Temperature at which development rate is maximum
Nonlinear	T_{max}	Upper developmental threshold or upper lethal temperature	the temperature at and above which the rate of development begins to decrease or stop
Nonlinear	Intrinsic optimum temperature (T_I)	Temperature at which rate controlling enzyme activity is maximal (thermodynamic model estimate)	Best temperature for an insect to survive; Reflects the adaption of an insect species to its thermal environment

* T_{min} and k can be estimated from logistic model as well (Davidson).

simulation that addresses individual variability in reproduction and is applicable for predicting reproduction under field variable temperatures and plant growth stages has been recently developed (Ma, 2010).

Phenology models help predict the time of events in an insects' development and are important analytical tools for predicting, evaluating and understanding the dynamics of pest populations in agro-ecosystems under a variety of environmental conditions. Accurate predictions, however, require accurate recording of the temperatures experienced by the organisms (Morgan, 1991) also the duration of development (Danks, 2000).

Degree-day models (Higley *et al.*, 1986) have long been used as part of decision support systems (DSS) to help growers predict spray timing or begin pest scouting (Welch *et al.*, 1978). Phenology models are also used as one component of risk analysis for predicting exotic pest establishment (Baker, 1991; Jarvis and Baker, 2001). A well-known example is the DYMEX modeling package (Su *et al.*, 2002; Yonow *et al.*, 2004; Stephens and Dentener, 2005). Some other modeling packages, for example CLIMEX although not strictly phenology models may also use some developmental requirements for risk assessment (Sutherst *et al.*, 1991, 1999, 2000). Another example is the web-based North Carolina State University-APHIS Plant Pest Forecast (NAPPFAST) modeling system that links daily climate and historical weather data with biological models to produce customized risk maps for phytosanitary risk assessments (Borchert and Magarey, 2005). Resources like the Crop Protection Compendium (CAB International, 2004) often summarize insect development, while the University of California Statewide IPM program lists development data for insects on their website (http://www.ipm.ucdavis.edu/MODELS) for use in DD models. An Insect Development Database containing the developmental requirements for over 500 insect species was created (Nietschke *et al.*, 2007). Insect Life Cycle Modeling software (ILCYM), a generic open source computer aided tool, facilitates the development of phenology models and prediction of pest activity in specific agroecologies (Sporleder *et al.*, 2009).

Life Tables and Population Models

Ecological life tables are one of the tools most useful in the study of insect population dynamics of insects having discrete generations. Such tables record a series of sequential measurements that reveal population change throughout the life cycle of a species in its natural environment. Conventionally, a life table is a systematic tabular presentation of survival and mortality in a population for a known cohort of individuals (Morris and Miller, 1954). Long term data from carefully planned population studies in which all of the relevant factors have been measured accurately are important for constructing population models which adequately relate to biological reality. The goal of life table analysis is to develop a population model which mimics reality. Apart from generating population estimates, this is best done by careful identification and measurement of the independent factors causing mortality such as parasitoids, predators, pathogens and weather factors.

From the life table studies it is possible to identify the key factor which is mainly responsible for increases and decreases in numbers from generation to generation

(Morris, 1963, Varley and Gradwell, 1970). A multiple regression approach involving all the survival components gives greater emphasis to the interaction between different age intervals (Mott, 1966). The equations for different mortalities are combined into a model to predict either the generation to generation changes in an insect population density or the average level about which these changes take place. The same analytical approaches used for insects having discrete generations are not applicable to insects with overlapping generations (Varley and Gradwell, 1970). Recently survival analysis was utilized to model both the development and survival of Russian wheat aphid (Ma and Bechinski, 2008). Ecological studies do not often lead to reliable forecast of the time and size of population peaks because of gaps in the ecological databases like short-range dispersal, overwintering behavior, colonization patterns and age-specific mortality including inter- and intra specific competition (Kogan and Turnipseed, 1987).

Pest Simulation Models and Decision Support Systems

Simulation models based on mathematical description of biological data as influenced by environment are more easily applied across locations and environments. Computer programs or software to run these models facilitates the practical application of these models in understanding population dynamics and dissemination of pest forecasts for timely pest management decisions (Coulson and Saunders, 1987) (Table 10.5). Simulation approach offers flexibility for testing, refinement, and sensitivity analysis and field validation of developed models over a wide range of environmental conditions. Thorough descriptions of cropping systems being managed or studied are needed to explain the interactions among pests, plants, and environment (Colbach, 2010). Systems models or other prediction schemes can be used with appropriate biological, environmental, economic, or other inputs to analyze the most effective management actions, based on acceptable control, sustainability, and assessment of economic or other risks (Strand, 2000).

Crop Growth Simulation Models Coupled with Pest Sub-routines

Crop system models can be used to generate information on the status of the crop, its pests, and its environment under different scenarios, including different management options. In practice, there are few examples of these models that include all the necessary components and can be used for practical decision making. However, a more practical approach has been the development of individual crop and pest components that can be analyzed at the same time to give information that can improve decisions.

Development of decision support system for agro-technology transfer (DSSAT 4 funded by USAID) allowed rapid assessment of several agricultural production systems around the world to facilitate decision-making at the farm and policy levels. The trend in development of crop system models is to go for modular approach (www.icasa.net). Development of stand-alone decision support systems for pest components could lead to their practical use. In developed countries, dynamic websites that include interactive models, GIS (Geographical Information System) based decision systems, real-time weather, and market information are being rapidly

Table 10.5: Selected Simulation Models in Use and their Outputs

Predictive Model	Inputs	Outputs	Reference
HEAPS, (*Helicoverpa armigera* and Punctigera Simulation), Australia	Adult movement, oviposition, development, survival and host phenology	Simulation of *H. armigera* and *H. punctigera* gridded populations, regional scale	Fitt *et al.,* 1995, Dillon and Fitt, 1990
MORPH (Methods of Research Practice in Horticulture), UK	Trap counts, weather	Timing of pest ac activity - cabbage root fly and carrot fly	Phelps *et al.,* 1999
ECAMON, Czech Republic	Daily weather data – minimum and maximum temperature, rain, daily sum of global radiation, precipitation, mean water vapor pressure and wind speed	European corn borer (ECB) activity - date of initiation, 25 per cent, 50 per cent, 75 per cent and termination(100 per cent) of each developmental stageusing a degree-day calculation method; Extent of environmental stresses on ECB-dryness, heavy rain, frost; ECB geographical distribution	Trnka *et al.,* 2007
RICEPEST, Phillipines	Pest and weather data	Yield loss simulation in production situations of tropical Asia	Willocquet *et al.,* 2002
SOPRA, Switzerland and Southern Germany	Hourly solar radiation, air and soil temperatures, Insect phenology	Prediction of age structure of populations for 8 major insect pests, local and regional scale	Samitez *et al.,* 2008
SIMLEP, Germany and Austria, West Poland, Slovenia	Daily weather data	Regional pest forecast of first occurrence of larval populations of Colorado potato beetle	Jorg *et al.,* 2007, Kos *et al.,* 2009
CIPRA (Computer Centre for Agricultural Pest Forecasting), Quebec	Air temperatures, RH, leaf wetness from a network of automatic weather stations (AWS)	Forecasts for 13 insects, two diseases, two storage disorders, and apple phenology	Burgeois *et al.,* 2008
DYMEX	Generic software, developmental, survival and fecundity rates, thresholds and weather	Daily simulations for timing and population size of cohorts	Sutherst, and Maywald, 2005
CLIMEX	Generic software, monthly weather data – minimum and maximum temperature and relative humidity, precipitation	Ecoclimatic index for development and survival of a species in specific climates and scenarios	Sutherst and Maywald, 1985
ILCYM (Insect life cycle modeling)	Generic software, developmental and life table data and temperatures	Insect developmental rates, thermal summation, pest distribution	Sporleder *et al.,* 2009

developed and made available on the Internet (www.effita.net) enabling farmers for real time use in crop management.

Conventional approaches of using empirical models for quantifying yield losses due to crop pests are limited in their scope and application since these equations are data-specific and insensitive to variable cropping and pest conditions. Crop growth models provide a physiologically based approach to simulate pest damage and crop interactions. There have been many efforts to use crop growth models to simulate the effect of pest damage on crop growth and yield by linking the damage effect of pest population levels to physiological rate and state variables of these models. The insect pest-crop modelling has been discussed in detail by Boote *et al.* (1983); Coulson and Saunders (1987). A distribution delay model including attrition was applied to simulate population changes of rice leaf-folders. Bases on metabolic pool approach, leaf folder feeding and hence leaf mass losses to the rice plant were described with a generalised functional response model which is 'source' and 'sink' driven (Graf *et al.*, 1992). Furthermore, this model stresses the influence of adult migration and natural enemies on leaf-folder population dynamics, both significant and poorly investigated aspects of the leaf-folder life cycle. Later, a generic approach to simulate the damage effects of single or multiple pests attempted using crop growth models such as CERES-rice (which is a part of DSSAT) in Philippines (Pinnschmidt *et al.*, 1995) and InfoCrop in India (Subash *et al.*, 2007; Reji *et al.*, 2008; Yadav and Subash Chander, 2010). Pest damage levels from field scouting reports can be entered and damage is applied to appropriate physiological coupling points within the crop growth model including leaf area index, stand density, intercepted light, photosynthesis, assimilate amount and translocation rate, growth of different plant organs and leaf senescence. Equations and algorithms were developed to describe competition among multiple pests and to link the computed total damage to the corresponding variables in the crop models. These approaches provide a basis to explore dynamic pest and crop interactions in determining pest management strategies which minimize yield losses.

Operational Pest Forecasting

Farmers are mainly interested in current disease and pest severity data, preferably for their localities to aid their decision-making in crop protection. Pest monitoring data along with complementary monitoring of weather data is crucial to run pest forecast models and make available forecasts for operational use. Weather measurements under field conditions from several geo-referenced sites in the crop cultivated regions additionally provides spatial information which can be used for generating pest forecast maps (Huang *et al.*, 2008). In Bayern (Germany) a measuring network of 116 field weather stations is used to estimate the development of pests in relation to weather requirements based on forecast models and computer-based DSS for near real time dissemination to farmers (Tischner, 2000). The results of crop and horticulture specific models and DSS are supplemented by field-monitoring data which then serve as the main input for the warning services and cost effectively disseminated through the Internet (Bugiani *et al.*, 1996; Jorg, 2000). A computerized national forecasting network in apple orchards transmits data from the field to system headquarters automatically. The national forecasting network in Turkey has been expanded and covered apple orchards in 34 provinces in 2006, using 115 electronic forecasting and warning stations (Atlamaz *et al.*, 2007).

An online tool has been developed for forecasting of mustard aphid (*Lipaphis erysimi*) from incidence data on mustard cultivars sown at staggered intervals at multiple locations over a period of 3 years. Multiple step-wise regression models were built between aphid incidence over time and composite weather variables with correlation coefficients as weights. Weather indices were computed for different weeks after sowing until the forecast was provided. Using SAS statistical software, models were devised for forecasting 1) crop age at first appearance of aphid on crop 2) crop age at highest aphid population (aphid counts on 10 cm of terminal shoot) in the season and 3) Peak aphid population on the crop in the season for different locations. The architecture of the online forecasting tool (Vinod Kumar *et al.*, 2012) is depicted in Figure 10.5.

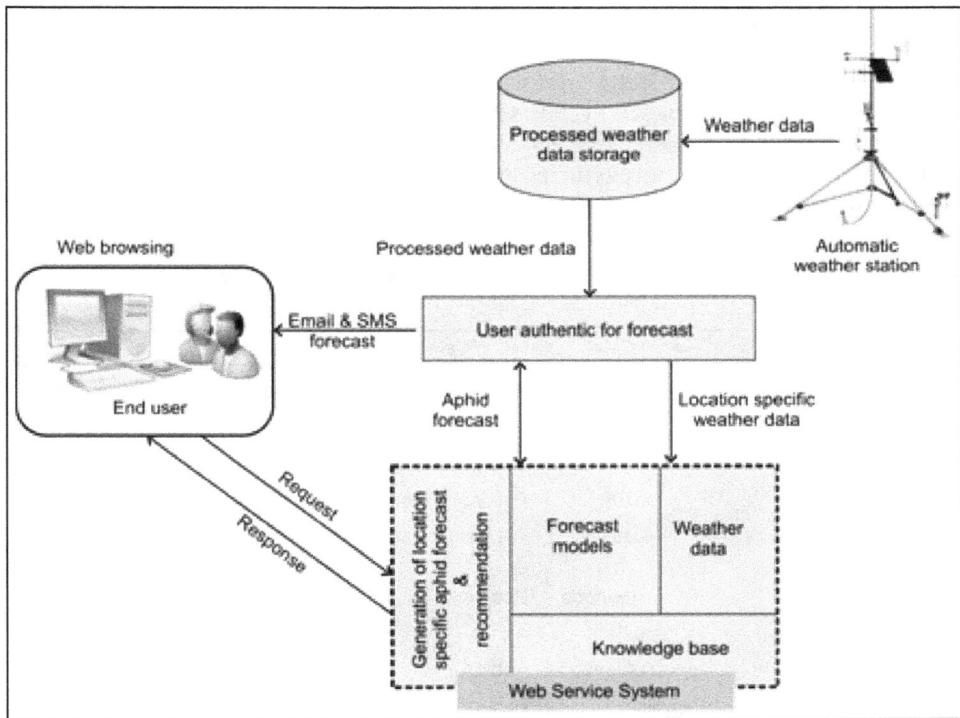

Figure 10.5: Online Mustard Aphid Forecast System (Vinod Kumar *et al.*, 2012).

'RB-Pred' a web-based server, was developed for forecasting leaf blast severity in rice based on the weather variables (Kaundal *et al.*, 2006). 'RB-Pred' predicts rice leaf blast severity (per cent) based on the weather parameters input by the user. It uses the regression module called support vector regression (SVR), a powerful machine learning technique called support vector machine (SVM). The SVM learns how to classify from a training set of feature vectors, whose expected outputs are already known. The training enables a binary classifying SVM to define a plane in the feature space, which optimally separates the training vectors of two classes. When a new feature vector is inputted, its class is predicted on the basis of which side of the plane it maps.

Conclusions

Pest monitoring is the foundation for issue of early warnings, development and validation of pest forecast models and decision support systems which are crucial for design and implementation of successful IPM programmes. Models are useful ways of synthesizing the available information and knowledge on population dynamics of pests in agro-ecosystems and natural habitats. However, any model needs to be tested and validated prior to use. Development of long-term monitoring on a spatial scale on crop-pest-weather relationships will narrow the gaps in knowledge required for reliable forecasts. In the last decade, two ICAR projects were funded by the World Bank and implemented with CRIDA as the consortium lead in the area of pest forecast research in key field crops such as rice, cotton, groundnut, sugarcane, pigeonpea and mustard. Significant contributions were made under the NATP project and useful leads are forthcoming under the ongoing NAIP project (Table 10.6). Computer-based systems have increased the speed and accuracy of forecasting as well as decreasing its costs. Recent developments in information and communication technology offer great scope for wide dissemination and use of pest forecasts. In the tropics, agro-ecosystems are characterized by greater crop

Table 10.6: Prediction Models Developed under the National Agricultural Technology Project (NATP) and National Agricultural Innovation Project of ICAR

Model/Tool	Crop	Pest	Reference
Life cycle modeling	Cotton	Mealybug	Prasad *et al.*, 2012
Life cycle modeling	Rice	Leaf folder	Padmavathi *et al.*, 2012
CART and NN models	Rice	Yellow stem borer	Amrender Kumar *et al.*, 2012
Regression	Cotton	Pink bollworm	Vennila *et al.*, 2011
Artificial neural network	Mustard	Powdery mildew	Ratna Raj Laxmi and Amrender Kumar, 2012
Regression	Pigeonpea	Pod borer	Vishwa Dhar *et al.*, 2008
Degree-day model	Mustard	Aphid	Chakravarthy and Gautam, 2004
Regression	Mustard	Aphid	Chattopadhyay *et al.*, 2005
Regression	Rice	Rice blast	Yella Reddy *et al.*, 2006
Epidemiological, Regression	Mustard	*Alternaria* blight	Chattopadhyay *et al.*, 2005
Epidemiological, regression	Mustard	Powdery mildew	Desai *et al.*, 2004
Online decision tools	Generic	Generic	http://www.crida.in/naip/comp4/dss_pest.html
Online forecast tool	Mustard	Aphid	Vinod Kumar *et al.*, 2012 (www.drmr.res.in/aphidforecast/models.html)
Online forecast tool	Rice	Blast	Kaundal *et al.*, 2006 (www.imtech.res.in/raghava/rbpred/)

diversity in small parcels of land and dynamically changing weather. Available generic simulation models need to be tested with location specific inputs for greater accuracy. In developing countries there is a strong need to establish functional networks for specific crop sectors with the major objective of pest forecasting through models and decision support systems.

References

Allen I C 1976. A modified sine wave method for calculating degree days. *Environmental Entomology* **5**: 388-396.

Anonymous 2009. Progress Report 2008 - Crop protection (Entomology, Plant pathology), All India Coordinated Rice Improvement Programme (ICAR), Directorate of Rice Research, Rajendranagar, Hyderabad, Andhra Pradesh (2009).

Atlamaz A, Zeki C and Uludag 2007. A. The importance of forecasting and warning systems in implementation of integrated pest management in apple orchards in Turkey. *EPPO Bulletin* **37**: 295-299.

Ayalew G, Sciarretta A, Baumgartner J, Ogol C and Lohr B 2008. Spatial distribution of diamondback moth, *Plutella xylostella* 1. (Lepidoptera: Plutellidae), at the field and the regional level in Ethiopia. *International Journal of Pest Management* **54**: 31-38.

Baker C R B 1991. The validation and use of a life-cycle simulation model for risk assessment of insect pests. *EPPO Bulletin* **21**: 615-622.

Birch L C 1957. The role of weather in determining the distribution and abundance of animals. In Cold Spring Harbor Symposia on *Quantitative Biology* **22**: 203-218. Cold Spring Harbor Laboratory Press.

Boote K J, Jones J W, Mishoe J W and Berger R. D 1983. Coupling pests to crop growth simulators to predict yield reductions. *Phytopathology* **73**: 1581-1587.

Borchert D M and Magarey R D 2005. A guide to the use of NAPPFAST./http://www.nappfast.org/usermanual/nappfast-manual.pdfS.

Bourgeois G, Plouffe D, Chouinard G, Beaudry N, Choquette D, Carisse O and DeEll J 2008. The apple Cipra network in Canada: using real-time weather information to forecast apple phenology, insects, diseases and physiological disorders. *Acta Horticulturae No: 803*.

Bowden J 1982. An analysis of factors affecting catches of insects in light traps. *Bulletin of Entomological Research* **72**: 535-556.

Bugiani R, Tiso R, Butturini A, Govoni P and Ponti I 1996. Forecasting models and warning services in Emilia-Romagna (Italy). *EPPO Bulletin* **26**: 595-603.

CAB International Crop Protection Compendium 2004. CAB International, Wallingford, UK.

Cannon R J C, Koerper D and Ashby S 2004. Gypsy moth, *Lymantria dispar*, outbreak in northeast London, 1995-2003. *International Journal of Pest Management* **50**: 259-273.

Chakravarty N V K and Gautam R D 2004. A degree-day based forewarning system for mustard aphid, *Lipaphis erysimi* (Kaltenbach). *Journal of Agrometeorology* **5**: 215-222.

Chattopadhyay C, Agrawal R, Kumar A, Bhar L M, Meena P D, Meena R L, Khan S A, Chattopadhyay A K, Awasthi R P, Singh S N, Chakravarthy N V K, Kumar A, Singh R B and Bhunia C K 2005a. Epidemiology and forecasting of *Alternaria* blight of oilseed *Brassica* inIndia - a case study. *Zeitschrift fürPflanzenkrankheiten und Pflanzenschutz (Journal of Plant Diseases and Protection)* **112**: 351-365.

Chattopadhyay C, Agrawal R, Kumar Amrender, Singh Y P, Roy, S K, Khan S A, Bhar L M, Chakravarthy N V K, Srivastava A, Patel B S, Srivastava B, Singh C P and Mehta S C 2005b. Forecasting of *Lipaphis erysimi* on oilseed *Brassicas* in India - a case study. *Crop Protection* **24**: 1042-1053.

Colbach N 2010. Modelling cropping system effects on crop pest dynamics: how to compromise between process analysis and decision aid. *Plant Science* **179**: 1-13.

Coulson R N and Saunders M C 1987. Computer-assisted decision making as applied to entomology. *Annual Review of Entomology* **32**: 415-437.

Curry G L, Feldman R M and Smith K C 1978. A stochastic model for a temperature-dependent population. *Theoretical Population Biology* **13**: 197-213.

Danks H V 2000. Measuring and reporting life-cycle duration in insects and arachnids. *European Journal of Entomology* **97**: 285-303.

Desai A G Chattopadhyay C Agrawal Ranjana Kumar A Meena R L Meena P D Sharma KC Srinivasa Rao M Prasad Y G and Ramakrishna Y S 2004. *Brassica juncea* powdery mildew epidemiology and weather- based forecasting models for India - a case study, *Zeitschrift für Pflanzenkrankheiten und Pflanzenschutz Journal of Plant Diseases and Protection* **111**: 429-438.

Dillon M L and Fitt G P 1990. HEAPS; a regional model of Heliothis population dynamics, pages 337-344 In: Proceedings of fifth Australian Cotton Conference 8-9 August, 1990, Broadbeach Queensland, Brisbane, Queensland, Australia: Australian Cotton Grower's Research Association.

Drake V A, Wang H K and Harman I T 2002. Insect monitoring radar: remote and network operation. *Computers and Electronics in Agriculture* **35**: 77-94.

El-sayed A M, Suckling D M, Wearing C H and Byers J 2006. A. Potential of mass trapping for long-term pest management and eradication of invasive species. *Journal of Economic Entomology* **99**: 1550-1564.

Feng H, Wu X, Wu B and Wu K 2009. Seasonal migration of *Helicoverpa armigera* (Lepidoptera: Noctuidae) over the bohai sea. *Journal of Economic Entomology* **102**: 95-104.

Fitt G P, Dillon M L and Hamilton J G 1995. Spatial dynamics of *Helicoverpa* populations in Australia: Simulation modeling and empirical studies of adult movement. *Computers and Electronics in Agriculture* **13**: 177-192.

Gallardo A, Ocete R, Lopez M.A, Maistrello L, Ortega F, Semedo A and Soria F J

2009. Forecasting the flight activity of *Lobesia botrana* (Denis and Schiffermüller) (Lepidoptera, Tortricidae) in Southwestern Spain. *Journal of Applied Entomology* **133**: 626-632.

Graf B, Lamb R, Heong K L and Fabellar L 1992. A simulation model for the population dynamics of rice leaf folders (Lepidoptera: Pyralidae) and their interactions with rice. *Journal of Applied Ecology* **29**: 558-570.

Gurrero A and Reddy G V P 2001. Optimum timing of insecticide applications against diamondback moth *Plutella xylostella* in cole crops using threshold catches in sex pheromone traps. *Pest Management Science* **57**: 90-94.

Gwadi K W, Dike M C, Amatobi C I 2006. Seasonal trend of flight activity of the pearl millet stemborer *Coniesta ignefusalis* (Lepidoptera: Pyralidae) as indicated by pheromone trap catches and its relationship with weather factors at samara, Nigeria. *International Journal of Tropical Insect Science* **26**: 41-47.

Higley L G, Pedigo L P and Ostlie K R 1986. DEGDAY: a program for calculating degree-days, and assumptions behind the degree-day approach. *Environmental Entomology* **15**: 999-1016.

Hopkinson R F and Soroka J J. Air trajectory model applied to an in-depth diagnosis of potential diamondback moth infestations on the Canadian prairies. *Agricultural and Forest Meteorology* **150**: 1-11.

Huang Y, Lan Y, Westbrook J K and Hoffmann W C 2008. Remote Sensing and GIS Applications for Precision Area-Wide Pest Management: Implications for Homeland Security. *The GeoJournal Library* **94**: 241-255.

Hultin E 1955. Influence of temperature on the rate of enzymatic processes. *Acta Chemica candinavica.,* 9, 1700 informations systems. In: J. Kroschel and L. Lacey [eds.], Integrated Pest Management for the Potato tuber moth *Phthorimaea operculella* (Zeller) - A potato pest of global importance. *Tropical Agriculture 20 - Advances in Crop Research 10*, Margraf Verlag, Weikersheim, Germany.

Ikemoto T 2005. Intrinsic optimum temperature for development of insects and mites. *Environmental Entomology* **34**: 1377-1387.

Jarvis C H and Baker R H 2001. A Risk assessment for non indigenous pests, I. Mapping the outputs of phenology models to assess the likelihood of establishment. *Diversity Distributions* **7**: 223-235.

Jennifer S O and Allen D C 2007. Monitoring populations of saddled prominent (Lepidoptera: Notodontidae) with pheromone-baited traps. *Journal of Economic Entomology* **100**: 335-342.

Johnson F H and Lewin I 1946. The growth rate of *E. Coli* in relation to temperature, quinine and coenzyme. *Journal of Cellular and Comparative Physiology* **28**: 47-75.

Jorg E 2000. Structure and organization of warning services in Rheinland-Pfalz (Germany). *EPPO Bulletin* **30**: 31-35.

Jorg E, Racca P, Preib U, Butturini A, Schmiedl J and Wojtowicz A 2007. Control of Colorado potato beetle with the simlep decision support system. *EPPO Bulletin* **37**: 353-358.

Kato M, Itioka T, Sakai S, Momose K, Yamane S and Hamid A 2000. A. and Inoue, T. Various population fluctuation patterns of light attaracted beetles in a tropical lowland dipterocarp forest in Sarawak. *Population Ecology* **42**: 97-104.

Kaundal R Kapoor A S and Raghava G P S 2006. Machine learning techniques in disease forecasting: a case study on rice blast prediction. *BMC Bioinformatics* **7**: 485.

Klueken A M, Hau B, Ulber B and Poehling H M 2009. Forecasting migration of cereal aphids (Hemiptera: Aphididae) in autumn and spring. *Journal of Applied Entomology* **133**: 328-344.

Knutson, Allen E, Muegge and Mark A 2010. A degree-day model initiated by pheromone trap captures for managing pecan nut casebearer (Lepidoptera: Pyralidae) in pecans. *Journal of Economic Entomology* **103**: 735-743.

Kogan M and Turnipseed S G 1987. Ecology and Management of soybean arthropods. *Annual Review of Entomology* **32**: 507-538.

Kumar Amrender, Prasad Y G Vennila S Vasantabhanu K Prabhakar M Padmakumari A P K and Katti G 2012. A comparative analysis of classification and regression tree (CART) and neural network (NN) models in prediction of rice yellow stem borer. 14th Annual Conference of Society of Statistics, Computer and Applications organised at the Department of Statistics, Saurashtra University, Rajkot during 24-26 February 2012.

Lewis T 1980. Britain's pest monitoring network for aphids and moths. *EPPO Bulletin* **10**: 39-46.

Liebhold A M and Tobin P C 2008. Population ecology of insect invasions and their management. *Annual Review of Entomology* **53**: 387-408.

Logan J A and Weber L A 1989. Population model design system (PMDS), user's guide. *Department of Entomology, Virginia Ploytechnic Institute and State University. Blackburg, VA.*

Logan J A, Wollkind D T, Hoyt J C and Tanigoshi L K 1976. An analytic model for description of temperature dependent rate phenomena in arthropods. *Environmental Entomology* **5**: 1130-1140.

Ma Z S 2010. Survival analysis approach to insect life table analysis and hypothesis testing: with particular reference to Russian wheat aphid (*Diuraphis noxia* (Mordvilko) populations. *Bulletin of Entomological Research* **100**: 315-324.

Ma Z S and Bechinski E J 2008. A survival-analysis-based simulation model for Russian wheat aphid population dynamics. *Ecological Modelling* **216**: 323-332.

Monobrullah M, Bharti P, Shankar U, Gupta R K, Srivastava K and Ahmad H 2007. Trap catches and seasonal incidence of *Spodoptera litura* on cauliflower and tomato. *Annals of Plant Protection Sciences* **15**: 73-76.

Morgan D 1991. The sensitivity of simulation models to temperature changes.

EPPO Bulletin **21**: 393-397.

Morris R F, Miller C A 1954. The development of life tables for the spruce budworm. *Canadian Journal of Zoology* **32**: 283-301.

Morris R.F Ed 1963. The dynamics of epidemic spruce budworm populations. *Memoirs of the Entomological Society of Canada* **31**: 332.

Mott D G 1966. The analysis of determination in population systems. *In: Systems Analysis in Ecology*, Chapter **7**: 179-194.

Nietschke B S, Magarey R D, Borchert D M, Calvin D D and Jones E 2007. A developmental database to support insect phenology models. *Crop Protection* **26**: 1444-1448.

Odiyo P O 1979. Forecasting infestations of a migrant pest: the African armyworm *Spodoptera exempta* (Walk.). *Philosophical Transactions of the Royal Society B: Biological Sciences* **287**: 403-413.

Padmavathi Chintalapati Gururaj Katti Sailaja, V Padmakumari, A P Jhansilakhsmi, V Prabhakar M and Prasad Y G 2012. Temperature thresholds and thermal requirements for the development of rice leaf folder, *Cnaphalocrocis medinalis. Journal of Insect Science* **13**: 1-14.

Phelps K, Collier R H, Reader R J and Finch S 1993. Monte Carlo simulation method for forecasting the timing of pest insect attacks. *Crop Protection* **12**: 335- 341.

Phelps K, Reader R J and Hinde C J 1999. HIPPO - flexible software for the construction, integration and distribution of biologically realistic models. *Aspects of Applied Biology* **55**: 81-88.

Pinnschmidt H O, Batchelor W D and Teng P S 1995. Simulation of multiple species pest damage in rice using CERES-rice. *Agricultural Systems* **48**: 193-222.

Prasad N V V S D, Mahalakshmi M S and Rao N H P 2008. Monitoring of cotton bollworms through pheromone traps and impact of abiotic factors on trap catch. *Journal of Entomological Research* **32**: 187-192.

Prasad Y G Prabhakar M Sreedevi G Ramachandra Rao G and Venkateswarlu B 2012. Effect of temperature on development, survival and reproduction of the mealybug, *Phenacoccus solenopsis* Tinsley (Hemiptera: Pseudococcidae) on cotton. *Crop Protection* **39**: 81-88.

Pratheepa M, Meena K,Subramaniam K R,Venugopalan R and Bheemanna H 2011. A decision tree analysis for predicting the occurrence of the pest, *Helicoverpa armigera* and its natural enemies on cotton based on economic threshold level. *Current Science* **100**: 242-246.

Raimondo S, Strazanac J S and Butler L 2004. Comparison of sampling techniques used in studying Lepidoptera population dynamics. *Environmental Entomology* **33**: 418-425.

Ratna Raj Laxmi and Amrender Kumar 2011. Forecasting of powdery mildew in mustard (*Brassica juncea*) crop using artificial neural networks approach, *Indian Journal of Agricultural Sciences* **81**: 855-860.

Regniere J 1984. A method of describing and using variability in development rates for the simulation of insect phenology. *Canadian Entomologist* **116**: 1367-1376.

Reji G and Chander S 2008. A degree-day simulation model for the population dynamics of the rice bug, *Leptocorisa acuta* (Thunb.). *Journal of Applied Entomology* **132**: 646-653.

Reji G, Chander S and Aggarwal P K 2008. Simulating rice stem borer, *Scirpophaga incertulas* Wlk., damage for developing decision support tools. *Crop Protection* **27**: 1194-1199.

Samietz J Graf B Höhn H Schaub L and Höpli H U 2007. Phenology modelling of major insect pests in fruit orchards from biological basics to decision support: the forecasting tool SOPRA, *Bulletin OEPP/EPPO Bulletin* **37**: 255-260.

Schoolfield R M, Sharpe P J and Magnuson C E 1981. Nonlinear regression of biological temperature-dependent rate models based on absolute reaction-rate theory. *Journal of Theoretical Biology* **88**: 719-731.

Shaffer P L 1983. Prediction of variation in development period of insects and mites reared at constant temperatures. *Environmental Entomology* **12**: 1012-1012.

Sharpe J H and DeMichele D W 1977. Reaction kinetics of poikilotherm development. *Journal of Theoretical Biology* **64**: 649-670.

Sharpe P J H, Curry G L, DeMichele D W and Coel C L 1977. Distribution model of organism development times. *Journal of Theoretical Biology* **66**: 21-28.

Sporleder M, Simon R., Gonzales J, Chavez D, Juarez H, De Mendiburu F and Krosche J 2009. ILCYM-Insect Life Cycle Modelling. A software package for developing temperature based insect phenologymodels with applications for regional and global risk assessments and mapping. International Potato Centre., Lima, Peru. pp 62.

Stephens A E A and Dentener P R 2005. *Thrips palmi*-potential survival and population growth in New Zealand. *NZ Plant Protection* **58**: 24-30.

Stinner R E, Butler G D, Bachelor J S and Tuttle C 1975. Simulation of temperature dependent development in population dynamics models. *Canadian Entomologist* **107**: 1167-1174.

Stinner R E, Gutierrez A P and Butler G D Jr 1974. An algorithm for temperature-dependent growth rate simulation. *Canadian Entomologist* **106**: 519-524.

Strand J F Some agrometeorological aspects of pest and disease management for the 21st century. *Agricultural and Forest Meteorology* **103**: 73-82.

Su J W and Fa S C 2002. Trapping effect of synthetic pheromone blends on two stem borers *Chilo syppressalis* and *Scirpophaga incertulas*. *CRRN-Chinese Rice Research Newsletter* **10**: 8-9.

Subash Chander, Kalra N and Aggarwal P K 2007. Development and application of crop growth simulation modelling in pest management. *Outlook on Agriculture* **36**: 63-70.

Sutherst R W, Maywald G F, Yonow T and Stevens P M 1999. CLIMEX: Predicting the Effects of Climate on Plants and Animals. *CSIRO Publishing*, Collingwood, Australia, 88pp.

Sutherst R W and Maywald G F 1985. A computerised system for matching climates in ecology, *Agriculture Ecosystems and Environment* **13**: 281-99.

Sutherst R W and Maywald G F 2005. A climate-model of the red imported fire ant, *Solenopsis invicta* Buren (Hymenoptera: Formicidae): implications for invasion of new regions, particularly Oceania, *Environmental Entomology* **34**: 317-335.

Sutherst RW, Maywald G F and Russell B L 2000. Estimating vulnerability under global change: modular modelling of pests. *Agriculture Ecosystems and Environment* **82**: 303-319.

Sutherst RW, Maywald G F and Bottomley W 1991. From CLIMEX to PESKY, a generic expert system for pest risk assessment. *Bulletin OEPP* **21**: 595-608.

Thomson L J, Macfadyen S and Hoffman A A 2010. Predicting the effect of climate change on natural enemies of agricultural pests. *Biological Control* **52**: 296-306.

Tischner H 2000. Instruments of the warning service for plant protection in Bayern (Germany). *EPPO Bulletin* **30**: 103-104.

Trnka M, Muska F, Semeradvoa D, Dubrovsky M, Kocmankova E and Zalud Z 2007. European corn borer life stage model: Regional estimates of pest development and spatial distribution under present and future climate. *Ecological Modelling* **207**: 61-84.

Varley G C and Gradwell G R 1970. Recent advances in insect population dynamics. *Annual Review of Entomology* **15**: 1-24.

Venkataraman S and Krishnan A 1992. Weather in the incidence and control of pests and diseases. *In:* Crops and Weather (Eds Venkataraman S and Krishnan A) Publications and Information Division, Indian Council of Agricultural Research, New Delhi. pp. 259-302.

Vennila S Meenu Agarwal Dharmendra Singh Prasenjit Pal and Biradar V K 2011. Approaches to weather based prediction of insects: a case study on cotton pink bollworm *Pectinophora gossypiella. Indian Journal of Plant Protection* **39**: 163-169.

Vinod Kumar Amrender Kumar and Chattopadhyay C 2012. Design and implementation of web-based aphid (*Lipaphis erysimi*) forecast system for oilseed *Brassicas. Indian Journal of Agricultural Sciences* **82**: 608-614.

Vishwa Dhar Singh S K Trivedi T P Das D K Choudhary R G and Kumar M 2008. Forecasting of *Helicoverpa armigera* infestation on long duration pigeonpea in central Uttar Pradesh. *Journal of Food Legumes* **21**: 189-192.

Wagner T L, Wu H I, Sharpe P J H and Coulson R N 1984b. Modeling distributions of insect development time: a literature review and application of the Weibull function. *Annals of the Entomological Society of America* **77**: 475-487.

Wagner T L, Wu H I, Feldman R M, Sharpe P J H and Coulson R N 1985. Multiple-cohort approach for simulating development of insect populations under variable temperatures. *Annals of the Entomological Society of America* **78**: 691-704.

Wagner T L, Wu H I, Sharpe P J H, Schoolfield R M and Coulson R N 1984a. Modelling insect development rates; a literature review application of a biophysical model, *Annals of the Entomological Society of America* **77**: 208-225.

Wall C, Garthwaite D G, Blood Smyth J A and Sherwood A 1987. The efficacy of sex-attractant monitoring for the pea moth, *Cydia nigricana,* in England, 1980-1985. *Annals of Applied Biology* **110**: 223-229.

Watt A D and Woiwod I P 1999. The effects of phonological asynchrony on population dynamics: analysis of fluctuations of British macrolepidopters. *Oikos* **87**: 411-416.

Welch S M, Croft B A and Michels M F 1978. PETE: an extension phenology modeling for management of muti-species pest complex. *Environmental Entomology* **7**: 487-494.

Willocquet L, Savary S, Fernandez L, Elazegui F A, Castilla N, Zhu D, Tang Q, Huang S, Lin X, Singh H. M and Srivastava R K 2002. Structure and validation of RICEPEST, a production situation-driven, crop growth model simulating rice yield response to multiple pest injuries for tropical Asia. *Ecological Modelling* **153**: 247-268.

Witzgall P, Kirsch P, Cork A 2010. Sex pheromones and their impact on pest management. *Journal of Chemical Ecology* **36**: 80-100.

Wolda H 1978. Fluctuations in abundance of tropical insects. *American Naturalist* **112**: 1017-1045.

Wolda H 1992. Trends in abundance of tropical forest insects. *Oecologia* **89**: 47-52.

Yadav D S and Subash Chander 2010. Simulation of rice plant hopper damage for developing pest management decision support tools. *Crop Protection* **29**: 267-276.

Yella Reddy D Prabhakar M Ramakrishna Y S Reddy C S Prasad Y G and Nagalaksmi T 2006. Dynamic cumulative weather based index for forewarning of rice blast. *Journal of Agrometeorology* **8**: 1-6.

Yonow T, Zalucki M P,Sutherst R W, Dominiak B C,Maywald G F, Maelzer D A and Kriticos D J 2004. Modelling the population dynamics of the queensland fruit fly, *Bactrocera (Dacus)tryoni*: a cohort-based approach incorporating the effects of weather. *Ecological Modeling* **17**: 39-30.

Zalucki M P and Furlong M J 2005. Forecasting *Helicoverpa* populations in Australia: A comparison of regression based models and a bioclimatic based modeling approach. *Insect Science* **12**: 45-46.

Zhu M, Song Y, Uhm K B, Turner R W, Lee J H and Roderick G K 2000. Simulation of the long range migration of brown planthopper, *Nilaparvata lugens* (sbu), by using boundary layer atmospheric model and the geographic information system. *Journal of Asia-Pacific Entomology* **3**: 25-32.

Zou L, Stout M J and Ring D R 2004. Degree-day models for emergence and development of the rice water weevil (Coleoptera: Curculionidae) in Southwestern Louisiana. *Environmental Entomology* **33**: 1541-1548.

Chapter 11

Recent Trends in the Analysis of Pesticide Residues by Mass Spectrometric Techniques

S K Raza and S Alam

Institute of Pesticide Formulation Technology (IPFT),
Gurgaon – 122 016, Haryana
E-mail: ipft@rediffmail.com

ABSTRACT

Analysis of pesticide residue in various environmental, food and biological matrices has achieved great concern due to increased awareness about the harmful effects of these pesticides. Newer innovations are taking place not only for the analysis of these pesticide residues but also for the sample preparation particularly their extraction from complex matrices, clean up and derivatization to make them amenable to GC or GC-MS analysis, which are key steps in successful detection of these residues at ppm and ppb levels. Recently, liquid chromatography coupled to mass spectrometry (LC-MS) and tandem mass spectrometry (LC-MS/MS) is finding huge application in the analysis of pesticide residues due to its capability to analyse polar and thermolabile compounds without much sample preparation and derivatization. This review gives an account of the recent trends in the analysis of pesticide residues in complex matrices using mass spectrometric techniques.

Keywords: Pesticides, Residues, Analysis, Mass spectrometry.

Introduction

Pesticides are substances or mixture of substances intended for preventing, destroying, repelling or mitigating any pest. Pesticides are a special kind of products for crop protection. Crop protection products in general protect plants from damaging influences such as weeds, diseases or insects. A pesticide is generally a chemical or biological agent (such as a virus, bacterium, antimicrobial or disinfectant) that through its effect deters, incapacitates, kills or otherwise discourages pests. Target pests can include insects, plant pathogens, weeds, molluscs, birds, mammals, fish, nematodes and microbes that destroy property, cause nuisance, spread disease or are vectors for disease. Although there are human benefits to the use of pesticides, some also have drawbacks, such as potential toxicity to humans and other animals. According to the Stockholm Convention on Persistent Organic Pollutants (POP), 9 of the 12 most dangerous and persistent organic chemicals are pesticides. Pesticides are categorized into four main substituent chemicals, *viz.*, herbicides; fungicides; insecticides and bactericides.

Pesticides have been used around the world for many years. In earlier times they were a protection against fungi and insect pests (Miller, 2002; Rao *et al.*, 2007). The great increase in the use of pesticides occurred with the development of new organic chemicals following World Wars I and II. In addition to chemicals for the control of fungi and insects, new developments were nematicides, herbicides, rodenticides, avicides, defoliants, wood preservatives, etc. The use of chemicals helped increase productivity, but caused great concern about their effect on human health and safety. On the other hand, chemicals did help tremendously from the standpoint of protecting against diseases that were carried by insects, especially mosquitoes. Adverse publicity has caused great concern about pesticides and this is especially so since our society has undergone great changes from an agricultural society to an industrial society and finally to a communications society. Unfortunately, publicity relating to the use of pesticides has seldom been balanced from the standpoint of the good and the bad. In fact, the communications media has and does usually stress the potential adverse effects of pesticides without reference to the good. This has caused concern on the part of advocates and the average person to the extent that it has placed heavy constraints on agriculture.

Pesticide residue refers to the pesticides that may remain on or in food after they are applied to food crops (Nic *et al.*, 2006). The levels of these residues in foods are often stipulated by regulatory bodies in many countries. Exposure of the general population to these residues most commonly occurs through consumption of treated food sources, or being in close contact to areas treated with pesticides such as farms or lawns around houses. Many of these chemical residues, especially derivatives of chlorinated pesticides, exhibit bioaccumulation, which could build up to harmful levels in the body as well as in the environment. Persistent chemicals can be magnified through the food chain, and have been detected in products ranging from meat, poultry, and fish, to vegetable oils, nuts, and various fruits and vegetables (Stephen *et al.*, 2011).

The human health protection from exposure to pesticides residues due to environment and food contamination has become a major concern due to increased usage of pesticides. Indeed over 1000 compounds may be applied to agricultural crops in order to control undesirable moulds, insects or weeds. To ensure the safety of food for consumers, numerous legislations such as codex directives have established maximum residue limits (MRLs) for pesticides in foodstuffs.

Pesticide Residue Analysis

Pesticide residue analysis in agricultural commodities and environmental matrices offers some peculiar features as compared to other organic trace analysis. These are: (i) wide range of analyses, with different polarities, solubility at different concentrations levels, may be determined in the same sample; (ii) wide range of commodities with different matrix effects in the determination of analysis due to different water and fat content and **biochemical composition**; (iii) Certified reference materials (CRMs) are not available.

Pesticide residue analyses are routinely carried out by means of multi-residue methods based on homogenization of the sample with an appropriate solvent, separation of the liquid portion of the sample from insoluble material, purification and clean up by florisil column followed by final chromatographic/ mass spectrometric determination step. Organic solvents commonly used to extract pesticide residues in fresh fruit and vegetables are acetonitrile, acetone, Petroleum Ether and diethyl ether etc. An extensive clean-up of organic extraction is necessary to reduce adverse effects of matrix interferences to the quantification of residues *e.g.* the masking of residue peaks by co-eluted matrix component, the occurrence of false positives and/or the inaccurate quantitation.

The complete analysis protocol involves the following steps:

1. Sample collection
2. Sample preparation and clean-up of the sample
3. Concentration/Derivatization
4. Analysis by mass spectrometry

Sample Collection

The difficulties of sampling biotic and abiotic materials for pesticide residues in tropical countries are exacerbated in areas remote from suitable storage facilities or from the analytical laboratories themselves. Any delay in preserving the sample or extracting the pesticide residues means that there is an increased risk of degradation of any residues present, with a corresponding increase in the uncertainty regarding the analytical results and their interpretation. If analysis of shorter lived compounds (such as organophosphates or carbamates) is required, then the risk of loss is great. However, with some pesticides (particularly the more persistent chlorinated pesticides and some herbicides), the risks of loss are less. The rate of loss for all types of compounds is greater under tropical rather than temperate conditions.

The construction of a comprehensive residue sampling programme is a huge subject and beyond the scope of this talk. It is not possible to define a sampling regime for all circumstances and the local conditions will need to be taken into account in each case.

The nature of the sampling exercise and the collection of the samples themselves require careful thought and planning. Samples taken in the wrong way or without due care can be misleading, resulting in incomplete or wrongly directed conclusions. General information on the properties and relative persistence of the different pesticide groups and their methods of application will help the investigator to determine the samples which need tobe taken, *e.g.* whether the pesticides used and the area and method of treatment will potentially affect biotic or abiotic factors (or both) and help in the development of an appropriate residue sampling programme. The exact nature of the study and the material of interest (soil, vegetation, insects, animal tissues, etc.) define the way in which the samples are taken and preserved prior to analysis.

Sample Preparation and Clean-Up of Sample

The primary aims of the sample preparation process are extracting maximum percentage of pesticide residues present in the sample, removing matrix interferences and increasing the sensitivity of analysis and recovery factor. The most commonly employed techniques for extracting pesticides are liquid–liquid extraction (LLE) and solid-phase extraction (SPE) (Kin *et al.*, 2009). However, conventional sample preparation techniques are time consuming, expensive and hazardous to health due to the use of high volume of potentially toxic solvents. Because of the disadvantages of conventional extraction techniques, recent trends focus on the solvent-free sample preparation methods or those employing less organic solvents. Some of the important sample preparation techniques as per the recent trends have been discussed below-

- ☆ QuEChERS Method
- ☆ Solid Supported Liquid/Liquid Extraction (SLE)
- ☆ Matrix Solid Phase Dispersion (MSPD)
- ☆ Single-drop microextraction(SDME)
- ☆ Solid phase micro extraction (SPME)

The sample preparation method chosen depends largely upon the food matrix and target list of pesticides, as well as existing successful applications. In addition, many useful modifications for the above sample preparation approaches can be found in the literature.

From an analytical perspective, older pesticides have more temperature stability than those recently developed, and are primarily detected by GC with MS, ECD, NPD, TSD or PFPD detectors. The latest generations of biodegradable pesticides are more thermally labile. These residues are frequently analyzed using L-/MS/MS to avoid hot injection temperatures that can degrade thermally labile compounds when using GC.

QuEChERS Method

The QuEChERS (**Qu**ick, **E**asy, **Ch**eap, **E**ffective, **R**ugged, **S**afe) method is designed for extraction of multiple pesticide residues from fruit, vegetables and other low fat foods. It uses bulk primary and secondary amine (PSA) SPE sorbent to remove matrix impurities. PSA removes fatty acids, sugars and other H-bonding matrix co-extractables. QuEChERS offers significant advantages over traditional homogenization techniques.

Solid Supported Liquid/Liquid Extraction (SLE)

This is an exceptional medium for Solid supported liquid/liquid extraction (SLE). Compared to conventional Liquid/Liquid extraction, SLE offers simple gravity flow, walk away operation, elimination of unwanted emulsions, reduced technique dependence and improved results and throughput.

Matrix Solid Phase Dispersion (MSPD)

MSPD combines a solid or viscous food sample with a specified ratio of bonded silica, often with the use of a mortar and pestle. The process ensures homogenous tissue disruption and creates a paste as the tissue adsorbs to the silica surface. The paste is packed into a standard SPE cartridge. Appropriate non-polar solvent schemes are then used to extract compounds of interest. Florisil columns are often used to assist in the removal of polar impurities. MSPD is effective for both low and high fat food matrices and is preferred over traditional homogenization or classical SPE techniques.

Single-Drop Microextraction (SDME)

SDME is based on the principle of distribution of the analytes between a microdrop of extraction solvent at the tipof a micro syringe, and an aqueous phase. A solvent microdrop is exposed to an aqueous sample, where the analyte is extracted into the drop. In this way, high enrichment factors are obtained owing to the high ratio of sample volume to organic phase volume. After extraction, the microdrop is retracted back into the microsyringe and injected into instruments such as GC-MS, GC, and HPLC for further analysis (Pakade *et al.*, 2010).

Solid-Phase Microextraction (SPME)

The technique was first developed in 1989 by Pawliszyn and co-workers, and has been commercialized by Supelco since 1993. Analytes are absorbed into an absorptionphase coated on a fused-silica fiber surface, and then desorbedeither in an injection port of a GC by thermal desorption, or bysolvent desorption for HPLC analysis. Adsorption follows the principle of equilibrium partitioning between sample molecules and absorption phase in a time-dependent manner. The solvent-free characteristic of SPME technology improves greatly the working environment for operators, and minimizes the volume of discarded toxic solvents (Chen *et al.*, 2010).

Concentration/Derivatization

The next step after sample preparation and clean up of the sample is its concentration and derivatization if required. The concentration of the sample is achieved either by nitrogen purging or by rotary vacuum evaporation. For nitrogen purging a slow stream of high purity Nitrogen Gas is passed through the solution carefully so as not to lose analyte during the purging. For rota evaporation, specially designed rota vapours are used to remove organic solvents from the extracted samples. Here again the evaporation has to be done carefully so that analytes are not lost.

A variety of derivatization methods are in use to make the analyte amenable to GC and GC-MS analysis. The most commonly used derivatizations are Silylation, Acylation and Methylation in various forms. Again Silylation is the most preferred method due to ease of preparation and easy availability of the silylating agents.

Analysis by Mass Spectrometry

Mass spectrometry (MS) coupled to Gas chromatography (GC) or Liquid chromatography (LC) has become the most sought after technique for the analysis of pesticide residues in various environmental, food and biological matrices. These technologies (GC-MS and LC-MS) enable us to identify unambiguously the components of a mixture without prior separation and at sensitivity compatible with the quantities of residue encountered. Mass spectrometry coupled to the separation techniques is widely used for the analysis of individual pesticide residue as also for multi-residue analysis. The main advantages of analysis by these mass spectrometric techniques include the high speed of analysis, elimination of the necessity for isolating minute quantities from a mixture and unambiguous identification of the eluted components as compared to the conventional GC or HPLC methods. Mass spectrometry allows direct determination of molecular weight of the compound and the number and kind of heteroatoms present in a molecule. Additionally, complete structural information of the molecule can be obtained by proper interpretation of spectrum obtained.

GC-MS and GC-MS/MS Analysis of Pesticide Residues

Gas chromatography coupled to mass spectrometry (GC-MS) and tandem mass spectrometry (GC-MS/MS) have been used extensively for the detection and analysis of various types of pesticide residues in environmental, food and biological matrices. One of the most important parameters when considering GC-MS methods of analysis is the choice of the ionization mode. Sample matrix and sample preparation procedures including clean-up also dictate selection of the ionization method due to presence of co-eluting pesticides or matrix components which can interfere in analysis. If pesticides are electron-capturing such as those pesticides which contain halogen, NO_2, or P ester groups then they will generally give an enhanced response with negative chemical ionization (NCI) in comparison to electron impact (EI) or positive chemical ionization (PCI) (Liapis *et al.*, 2003; Bailey and Belzer, 2007; Húšková *et al.*, 2009; Raina and Hall, 2009). The selection of ionization mode often depends upon whether the analysis is targeted for specific chemical classes or is a multiresidue analysis method. GC-MS of a wider range of triazines has also been done by GC-EI/MS (Albanis *et al.*, 1998; Zambonin and Palmisano, 2000; Jiang *et al.*, 2005; Gonçalves *et al.*, 2006; Nagaraju and Huang, 2007). Due to the large diversity in properties of pesticides analyzed by multiresidue analysis methods EI is more frequently used. However, for many halogenated pesticides, it does not often give the best sensitivity or selectivity. The clear advantage of EI is the availability of extensive libraries in full scan mode for confirmation of compound by library search matching.

Selected Ion Monitoring (SIM) is the method of choice for quantitative analysis of pesticides. In SIM mode, the peak area of the most abundant ion in the MS spectra is used for the quantitative analysis, and the peak area obtained from an additional one or two ionsis used for confirmation along with the ratio of ion responses and retention time match (Raina and Hall, 2009). The Quick Easy Cheap Effective Rugged and Safe multiresidue (QuEChERS) method has been validated for the extraction of 80 pesticides belonging to various chemical classes from various types of representative commodities with low lipid contents. A mixture of 38 pesticides amenable to gas chromatography (GC)was quantitatively recovered from spiked lemon, raisins, wheat flour and cucumber, and determined using gas chromatography-tandem mass spectrometry (GC-MS/MS) technique (Paya *et al.*, 2007). The pesticides chosen included many of the most frequently detected ones and/or those that are most often found to violate the maximum residue limit (MRL) in food samples, some compounds that have only recently been introduced, as well as a few other miscellaneous compounds. The method employed involved initial extraction in a water/acetonitrile system, an extraction/partitioning step after the addition of salt, and a cleanup step utilizing dispersive solid-phase extraction (DSPE); this combination ensured that it was a rapid, simple and cost-effective procedure. Many researchers reported QuEChERS sample preparation in combination with mass spectrometric detection (Lehotay, 2005; Brown *et al.*, 2011, Liu *et al.*, 2011; Walorczyk *et al.*, 2011; Xu *et al.*, 2011). In addition to the QuEChERS technique, other solvent less or less solvent consuming sample preparation techniques in combination with Gas chromatography or liquid chromatography mass spectrometric detection are also getting equal importance in pesticide residue analysis. Bagheri *et al.* (2012) developed a simple, rapid and sensitive method based on microextraction in packed syringe (MEPS), in combination with gas chromatography-mass spectrometry (GC-MS). Li *et al.* (2012) established a method based on disposable pipet extraction (DPX) sample cleanup and gas chromatography with mass spectrometric detection by selected ion monitoring (GC/MS-SIM) for 58 targeted pesticide residues in soybean, mung bean, adzuki bean and black bean. A brief review on recent trends in pesticide residue analysis has been presented in Table 11.1.

LC-MS and LC-MS/MS Analysis of Pesticide Residues

LC-MS and LC-MS/MS continue to gain popularity in use for pesticide analysis with most applications focused on non-GC amenable, thermolabile, polar and non-volatile pesticides. Some chemical classes such as phenoxyacids, triazines, OPs, chloroacetanilides, and pyrethroids can be analyzed by both GC-MS and LC-MS. For phenoxyacid herbicides and carbamatesLC-MS and LC-MS/MS is regarded as more favourable as it does not require a derivatization step prior to analysis. The use of LC-MS/MS over GC/MS for these chemical classes may also be done in order to achieve reduced analysis time by utilizing a multiresidue LC-MS/MS method covering a range of target pesticides from different chemical classes. However, the key reason for choosing LC-MS/MS over GC-MS is the need to deal with more polar chemical classes of pesticides and increasingly for the simultaneous analysis of their transformation products.

Transformation products are often more polar and less volatile than their parent compounds and generally have poor chromatographic performance on nonpolar GC columns or are thermolabile. Transformation products many also require

Table 11.1: Combination of Solvent Free or Less Solvent Consuming Sample Preparation Techniques and Mass Spectrometric Analytical Techniques

Matrices	Sample Preparation	Analytical Techniques	Detection Limit	References
Water	SDME	GC–MS	0.010–0.73 mg/L	Lambropoulou et al., 2007
Fish	SDME	GC–MS	0.5 ng/g	Shrivas et al., 2008
Rice farm water and river water	SDME	GC–MS	0.015–0.4 mg/L	Bagheri et al., 2005
Vegetables	SDME	GC–MS	0.05–0.2 ng/mL	Zhang et al., 2008
Water	SDME	GC–MS	3–35 ng/L	Saraji et al., 2008
Water	SDME	GC–MS	1.2–7 ng/L	Saraji et al., 2008
Water	SDME	GC–MS	0.022–0.10 mg/L	Cortada et al., 2009
Wastewater	SDME	GC–MS	0.3–11.4 ng/L	Chanbasha et al., 2007
Grapes and apples	SDME	GC–MS	0.0003–0.007 mg/g	Amvrazi et al., 2009
Surface and waste water	SPE	LC/MS/MS	0.3 and 30 ng/L	Rodil et al., 2009
River water	On-line SPM	LC–ESI-MS/MS	50 ng/L	Kuster et al., 2008
Human urinary	Automated solid-phase extraction	GC/MS/MS	50–170 pg/mL	Hemakanthi De Alwis et al., 2006
Human urinary	SPE	GC/MS/MS	0.1–1.0_g/L	Schindler et al., 2009
Sewage sample	SPE	GC/MS	1–7 pg/g	Basheer et al., 2006
Sewage sludge samples	HS-SPME	GC/MS	21–93 pg/g	Tahboub et al., 2005
Sewage sludge samples	HFM-SPME	GC/MS	10–67 pg/g	Tahboub et al., 2005
Aquaculture seawater samples	SPME	GC/MS/MS	1.0–600 pg/mL	García-Rodríguez et al., 2008
Textiles	SPME	GC/MS	0.01–55g/L	Zhu et al., 2009
Tap water	SBSE	GC/MS	0.1–10.7 ng/L	Leon et al., 2006
Aqueous samples	SBSE	LTM–GC–MS	0.058–9.4 ng/L	Ochiai et al., 2006
Brazilian sugarcane juice	SBSE	GC–MS (SIM)	0.002–0.71g/L	Zuin et al., 2006
Saffron	SBSE	GC/MS/MS	0.06–0.56g/kg	Maggia et al., 2008

derivatization to make them GC amenable for some of these chemical classes. Even for LC-MS/MS methods the large difference in polarity between parent pesticide and transformation products may require different separation conditions or ion source (mode) for adequate sensitivity making development of simultaneous methods challenging. The pesticide residue analysis by LC-MS/MS has focused on systems with atmospheric pressure ionization (API) either atmospheric pressure chemical ionization (APCI) or electrospray ionization (ESI) in positive or negative ion mode. Both APCI and ESI have been used for multiresidue methods with ESI⁺being the most popular technique.

Liquid chromatography-mass spectrometry (LC-MS) is rapidly becoming a routine techniquefor efficient trace analysis of polar pesticides invarious types of samples. In comparison to gas chromatography-massspectrometry (GC-MS) and ultraviolet (UV) detection, LC-MS considerably simplifies cleanup procedures, reducing both time of analysis and method development time (Hogendoorn and van Zoonen, 2000).

More than 280 pesticide residues including difficult polar species have been analysed by using ultra high pressure LC-MS/MS technique showing excellent peak shapes and retention on a 3μm C18 column (Wittery, 2012).

A comparison of GC-MS or GC-MS/MS with EI to LC/MS/MS has been reviewed for a large number of compounds and suggests for most pesticides other than organochlorines that LC/MS/MS can provide lower detection limits (Alder *et al.*, 2006; Pihlström *et al.*, 2007; Paya *et al.*, 2007; Lambropoulou *et al.*, 2007). However, lower or comparable detection limits have also been found for chloracetanilides (metolachlor, acetochlor, alachlor) and selected triazines by GC/MS or GC/MS/ MS with EI relative to LC/APCI-MS/MS (Dagnac *et al.*, 2005) or LC/ESI-MS/MS (Freitas *et al.*, 2004).

Conclusion

Food and environment safety are the topics of great interest globally. With recent food contamination issues in a wide range of commodities, ensuring the quality of our food supply is becoming increasingly important. Pesticide residue content is one area of concern. With recent advances in mass spectrometric techniques, this technique is quickly gaining acceptance for pesticide residue testing. While GC-MS and GC-MS/MS are the methods of choice for volatile, non-polar and non-thermolabile compounds, LC-MS and LC-MS/MS can be used to simultaneously monitor non-volatile, polar and thermolabile contaminants including those difficult to detect by GC-MS. Both these approaches allow for a faster, more complete picture of pesticide residues. The use of selected ion monitoring (SIM) in MS mode and multiple reaction monitoring (MRM) in MS/MS mode further permits identification of the target pesticides with greater sensitivity and specificity.

References

Albanis T A, Hela D G, Sakellarides T M and Konstantinou I K 1998. Monitoring of pesticide residues and their metabolites in surface and underground waters of Imathia (N. Greece) by means of solid-phase extraction disks and gas chromatography. *Journal of Chromatography A* **823**: 59-71.

Alder L, Gruelich K, Kempe G and Vieth B 2006. Residue analysis of 500 high priority pesticides: Better by GC–MS or LC–MS/MS. *Mass Spectrometry Reviews* **25**: 838-865.

Amvrazi E.G. and Tsiropoulos N G 2009. Chemometric study and optimization of extraction parameters in single-drop microextraction for the determination of multiclass pesticide residues in grapes and apples by gas chromatography mass spectrometry. *Journal of Chromatography A* **1216**: 7630-7638.

Bagheri H, Alipour N and Ayazi Z 2012. Multiresidue determination of pesticides from aquatic media using polyaniline nanowires network as highly efficient sorbent for microextraction in packed syringe. *Analytica Chimica Acta* **740**: 43-49.

Bagheri H and Khalilian F 2005. Immersed solvent microextraction and gas chromatography–mass spectrometric detection of *s*-triazine herbicides in aquatic media. *Analytica Chimica Acta* **537**: 81-87.

Bailey (Raina) R and Belzer W 2007. Large Volume Cold On-Column Injection for Gas Chromatography–Negative Chemical Ionization–Mass Spectrometry Analysis of Selected Pesticides in Air Samples. *Journal of Agricultural and Food Chemistry* **55**: 1150-1155.

Basheer C, Alnedhary A A, MadhavaRao B S, Valliyaveettil S and Lee H K 2006. Development and Application of Porous Membrane-Protected Carbon Nanotube Micro-Solid-Phase Extraction Combined with Gas Chromatography/ Mass Spectrometry. *Analytical Chemistry* **78**: 2853-2858.

Brown A N, Cook J M, Hammack W T, Stepp J S, Pelt J V and Gerard G 2011. Analysis of Pesticides Residues in Fresh Produce Using Buffered Acetonitrile Extraction and Aminopropyl Cleanup with Gas Chromatography/Triple Quadrupole Mass Spectrometry, Liquid Chromatography/Triple Quadrupole Mass Spectrometry, Gas Chromatography/Ion Trap Detector Mass Spectrometry, and GC with a Halogen-Specific Detector. *Journal of AOAC International* **94**: 931-941.

Chanbasha B, Anass A A, MadhavaRao B S and Lee H K 2007. Determination of organophosphorous pesticides in wastewater samples using binary-solvent liquid-phase microextraction and solid-phase microextraction: A comparative study. *Analytica Chimica Acta* **605**: 147-152.

Chen J, Duan C and Guan Y 2010. Sorptive extraction techniques in sample preparation for organophosphorus pesticides in complex matrices. *Journal of Chromatography B* **878**: 1216-1225.

Cortada C, Vidal L, Tejada S, Romo A and Canals A 2009. Determination of organochlorine pesticides in complex matrices by single-drop microextraction coupled to gas chromatography–mass spectrometry. *Analytica Chimica Acta* **638**: 29-35.

García-Rodríguez D, Antonia M C, Rosa A, Lorenzo F and Fernández R C 2008. Determination of trace levels of aquaculture chemotherapeutants in seawater samples by SPME-GC-MS/MS. *Journal of Separation Science* **31:** 2882-2890.

Gomides Freitas L, Gotz C W, Ruff M, Singer H P and Muller S R 2004. Quantification of the new triketone herbicides, sulcotrione and mesotrione, and other important herbicides and metabolites, at the ng/l level in surface waters using liquid chromatography–tandem mass spectrometry. *Journal of Chromatography A* **1028:** 277-286.

Gonçalves C, Carvalho J J, Azenha M A and Alpendurada M F 2006. Optimization of supercritical fluid extraction of pesticide residues in soil by means of central composite design and analysis by gas chromatography–tandem mass spectrometry. *Journal of Chromatography A* **1110:** 6-14.

Goodman W 2007. The application of GC/MS to the analysis of pesticides in foodstuffs, Perkin Elmer Application Notes.

Hemakanthi De Alwis G K, Needham L L and Barr D B 2006. Measurement of human urinary organophosphate pesticide metabolites by automated solid-phase extraction, post extraction derivatization, and gas chromatography–tandem mass spectrometry. *Journal of Chromatography B* **843:** 34-41.

Hogendoorn E and van Zoonen P 2000. Recent and future developments of liquid chromatography in pesticide trace analysis. *Journal of Chromatography A* **892:** 435-453.

Húskova R, Matisova E, Švorc L, Mocák J and Kirchner M 2009. Comparison of negative chemical ionization and electron impact ionization in gas chromatography–mass spectrometry of endocrine disrupting pesticides. *Journal of Chromatography A* **1216:** 4927-4932.

Jiang H, Adams C D and Koffskey W 2005. Determination of chloro-*s*-triazines including didealkylatrazine using solid-phase extraction coupled with gas chromatography–mass spectrometry. *Journal of Chromatography A* **1064:** 219-226.

Kin C M and Haut T G 2009. Comparison of HS-SDME with SPME and SPE for the determination of eight organochlorine and organophosphorus pesticide residues in food matrices. *Journal of Chromatography* **47:** 694-699.

Kuster M, López de Alda M J, Barata C, Raldúa D and Barceló D 2008. Analysis of 17 polar to semi-polar pesticides in the Ebro river delta during the main growing season of rice by automated on-line solid-phase extraction-liquid chromatography–tandem mass spectrometry. *Talanta* **75:** 390-401.

Lambropoulou A D and Albanis A T 2007. Methods of sample preparation for determination of pesticide residues in food matrices by chromatography–mass spectrometry-based techniques: a review. *Analytical and Bioanalytical Chemistry.* **389:** 1663-1683.

Lehotay S J, Kok A, Hiemstra M and Bodegraven P 2005. Validation of a Fast and Easy Method for the Determination of Residues from 229 Pesticides in Fruits and Vegetables Using Gas and Liquid Chromatography and Mass Spectrometric Detection. *Journal of AOAC International* **88**: 595-614.

Leon V M, Llorca-Porcel J, Alvarez B, Cobollo M A, Munoz S and Valor I 2006. Analysis of 35 priority semivolatile compounds in water by stir bar sorptive extraction–thermal desorption–gas chromatography–mass spectrometry: Part II: Method validation. *Analytica Chimica Acta* **558**: 261-266.

Li Z, Li Y, Liu X, Li X, Zhou L and Pan C 2012. Multiresidue analysis of 58 pesticides in bean products by disposable pipet extraction (DPX) cleanup and gas chromatography-Mass spectrometry determination. *Journal of Agricultural and Food Chemistry* **60**: 4788-4798.

Liapis K S, Aplada-Sarlis P and Kyriakidis N V 2003. Rapid multi-residue method for the determination of azinphos methyl, bromopropylate, chlorpyrifos, dimethoate, parathion methyl and phosalone in apricots and peaches by using negative chemical ionization ion trap technology. *Journal of Chromatography A* **996**: 181-187.

Liu G, Rong L, Guo B, Zhang M, Li S, Wu Q, Chen J, Chen B and Yao S 2011. Development of an improved method to extract pesticide residues in foods using acetontrile with magnesium sulfate and chloroform. *Journal of Chromatography A* **1218**: 1429-1436.

Maggi L, Carmonaa M, del Campoa C P, Zalacaina A, de Mendozab J H, Mocholí F A and Alonso G L 2008. Multi-residue contaminants and pollutants analysis in saffron spice by stir bar sorptive extraction and gas chromatography–ion trap tandem mass spectrometry. *Journal of Chromatography A* **1209**: 55-60.

Miller G T 2002. *Living in the Environment* (12th Ed.). Belmont: Wadsworth/ Thomson Learning.

Nagaraju, D and Huang S D 2007. Determination of triazine herbicides in aqueous samples by dispersive liquid–liquid microextraction with gas chromatography–ion trap mass spectrometry. *Journal of Chromatography A* **1161**: 89-97.

Nic M, Jirat J and Kosata B eds 2006. Pesticide Residue. *IUPAC Compendium of Chemical Terminology* (Online ed.).

Ochiai N, Sasamoto K, Kandaa H and Nakamura S 2006. Fast screening of pesticide multiresidues in aqueous samples by dual stir bar sorptive extraction-thermal desorption-low thermal mass gas chromatography–mass spectrometry. *Journal of Chromatography A* **1130**: 83-90.

Pakade Y B and Tewary D K 2010. Development and applications of single-drop microextraction for pesticide residue analysis: A review. *Journal of Separation Science* **33**: 3683–3691.

Payá P, Anastassiades M, Mack D Sigalova I, Tasdelen B, Oliva J and Barba A 2007. Analysis of pesticide residues using the Quick Easy Cheap Effective Rugged and Safe (QuEChERS) pesticide multiresidue method in combination with gas and liquid chromatography and tandem mass spectrometric detection. *Analytical and Bioanalytical Chemistry* **389**: 1697-1714.

Pihlstrom T, Blomkvist G, Friman P, Pagard U and Osterdahl B G 2007. Analysis of pesticide residues in fruit and vegetables with ethyl acetate extraction using gas and liquid chromatography with tandem mass spectrometric detection. *Analytical and Bioanalytical Chemistry* **389:** 1773-1789.

Raina R and Hall P 2008. Comparison of Gas Chromatography-Mass Spectrometry and Gas Chromatography Tandem Mass Spectrometrywith Electron Ionization and Negative-Ion Chemical Ionization for Analyses of Pesticides at Trace Levelsin Atmospheric Samples. *Analytical Chemistry Insights* **3:** 111-125.

Rao G V R, Rupela O P, Rao V R and Reddy Y V R 2007. Role of biopesticides in crop protection: present status and future prospects. *Indian Journal of Plant Protection* **35:** 1-9.

Rodil R, Quintanab J B, López-Mahíaa P, Muniategui-Lorenzoa S and Prada-Rodríguez D 2009. Multi-residue analytical method for the determination of emerging pollutants in water by solid-phase extraction and liquid chromatography-tandem mass spectrometry. *Journal of Chromatography A* **1216:** 2958-2969.

Saraji M and Bahman F 2008. Application of single-drop microextraction combined with in-microvialderivatization for determination of acidic herbicides in water samples by gas chromatography–mass spectrometry. *Journal of Chromatography A* **1178:** 17-23.

Saraji M and Esteki N 2008. Analysis of carbamate pesticides in water samples using single-drop microextraction and gas chromatography-mass spectrometry. *Analytical and Bioanalytical Chemistry* **391:** 1091-1100.

Schindler B K, Förster K and Angerer J 2009. Determination of human urinary organophosphate flame retardant metabolites by solid-phase extraction and gas chromatography–tandem mass spectrometry. *Journal of Chromatography B* **877:** 375-381.

Shrivas K and Wu H F 2008. Ultrasonication followed by single-drop microextraction combined with GC/MS for rapid determination of organochlorine pesticides from fish. *Journal of Separation Science* **31:** 380-386.

Stephen W C C and Benedict L S C 2011. Determination of organ chlorine pesticide residues in fatty foods: A critical review on the analytical methods and their testing capabilities. *Journal of Chromatography A* **1218:** 5555-5567.

Tahboub Y R, Zaater M F and Al-Talla Z A 2005. Determination of the limits of identification and quantitation of selected organochlorine and organophosphorous pesticide residues in surface water by full-scan gas chromatography/mass spectrometry. *Journal of Chromatography A* **1098:** 150-155.

Walorczyk S, Drovdvyñski D and Gnusowski B 2011. Multiresidue determination of 160 pesticides in wines employing mixed-mode dispersive-solid phase extraction and gas chromatography–tandem mass spectrometry. *Talanta* **85:** 1856-1870.

Xu R, Wu J, Liu Y, Zhao R, Chen B, Yang M and Chen J 2011. Analysis of pesticide residues using the Quick Easy Cheap Effective Rugged and Safe (QuEChERS) pesticide multiresidue method in traditional Chinese medicine by gas chromatography with electron capture detection. *Chemosphere* **84:** 908-912.

Zambonin C G and Palmisano F 2000. Determination of triazines in soil leachates by solid-phase microextraction coupled to gas chromatography–mass spectrometry. *Journal of Chromatography A* **874:** 247-255.

Zhang M, Huang J, Wei C, Yu B, Yang X and Chen X 2008. Mixed liquids for single-drop microextraction of organochlorine pesticides in vegetables. *Talanta* **74:** 599-604.

Zhu F, Ruan W H, He M H, Zeng F, Luan T G, Tong Y X, Lu T B and Ouyang G F 2009. Application of solid-phase microextraction for the determination of organophosphorus pesticides in textiles by gas chromatography with mass spectrometry. *Analytica Chimica Acta* **650:** 202-206.

Zuin V G, Schellin M, Montero L, Yariwake J H, Augusto F and Poppb P 2006. Comparison of stir bar sorptive extraction and membrane-assisted solvent extraction as enrichment techniques for the determination of pesticide and benzo[a]pyrene residues in Brazilian sugarcane juice. *Journal of Chromatography A* **1114:** 180-187.

Chapter 12

Electron Beam Irradiaton: A Novel Disinfestation Technology of 21st Century for Phytosanitary Purposes

Shashi Bhalla[1]*, K Srinivasan[1],
J Dwivedi[2] and Arun Sharma[3]

[1]ICAR-National Bureau of Plant Genetic Resources, New Delhi – 110 012
[2]Raja Ramanna Centre for Advanced Technology, Indore – 452 013
[3]Bhabha Atomic Research Centre, Mumbai – 400 085
*E-mail: shashi.bhalla@icar.gov.in

ABSTRACT

In view of the environmental and health hazards posed by toxic chemicals used for crop protection, there is a need for alternative ecofriendly strategies for controlling the pests. Irradiation as a phytosanitary measure, has gained increasing acceptance in recent years. It has been recognized as a versatile treatment with broad spectrum activity against different pests. Significant contributions have been made worldwide using radiations for sanitary and phytosanitary purposes. It is an approved technology of International Plant Protection Convention for phytosanitary purposes and appears to be a viable, non-chemical and residue-free strategy. Irradiation promises rapid, efficacious quarantine treatment for many pests with minimal damage to commodity quality. Internationally, now new radiation generating sources as Electron beam are being explored to meet import standards of quality and quarantine. Electrons from accelerators present fewer health hazards than does the use of gamma rays, because electron beams are directional and less penetrating and have no hazard of radioactive isotopes. Electron beams have been reported to effectively disinfest grains, pulses, spices, etc. and have a great potential for use in the disinfestation of food grains with considerably less quality deterioration than other disinfestation techniques. Our results also confirm that electron beam irradiation is an effective method for the management of Callosobruchus maculatus *infesting mungbean. Nevetheless, there are many avenues which merit further exploration. The adoption of particular treatment depends on the nature of pest and commodity to be treated, its level of effectiveness, convenience and ease of use and economic considerations. These measures need to be used in developing the disinfestation protocols to meet the international obligations under the WTO regime.*

Keywords: *Electron beam, Irradiation, Disinfestation, Phytosanitary.*

The liberalized trade in WTO regime has increased the flow of agricultural commodities across different countries. A large number of plant samples are also being exchanged throughout the world for crop improvement programs. These exchanges always involve the risk of inadvertent introduction of plant pests along with the movement of planting material, which may prove disastrous as evident from several examples of epidemics in the history. Such pests, if established may prove more devastating in the new geographical areas in the absence of natural enemies and competitors. The risk of introduction of exotic pests could be minimized by undertaking quarantine of the plant material under exchange.

At present, India, imports agricultural commodities including germplasm as per the provisions of Plant Quarantine (Regulation of Import into India) Order 2003 promulgated by the Ministry of Agriculture, Government of India, which came into force with effect from April 1, 2004. Plant quarantine has important role in safe exchange of plant material, which is the legislative measure to regulate the introduction of planting material to prevent the inadvertent entry/establishment of the exotic pests into the country and so also the spread of the pests, already established, to other parts of the country. At national level, Department of Plant Protection Quarantine and Storage, Ministry of Agriculture, Government of India deals with the quarantine of bulk material, both for sowing and consumption purposes. The National Bureau of Plant Genetic Resources is the nodal organization with the authority to issue Import Permit and Phytosanitary Certificate and undertake quarantine processing of germplasm including transgenic planting material meant for research purposes. Every year about 70,000 samples are being imported into the country (India) for research purposes, which require quarantine. Quarantine processing involves inspection of the material for detection of pests and salvaging of the infested/infected material through quarantine/phytosanitary treatments. Quarantine/phytosanitary treatments demand a very high level of security as the pest tolerance in quarantine is zero. It is a measure that lowers the pest risk by destroying/eradicating a pest, thus preventing its dispersal to new areas. The treatment should be effective to give 100 per cent kill of the target pest without affecting the germination of the planting material or the acceptability of the commodity. However, practically it is not possible and hence, the concept of Probit 9 has been accepted in quarantine.

Fumigation with methyl bromide (MB) has been the most widely applied management practice for insect pests. MB, a potent quarantine fumigant being used world over is being restricted globally because of the problems related of toxic residues in food grains, development of insect resistance and adverse environmental impacts. It has been considered as an ozone depleting substance under the Montreal Protocol and has to be phased out in the near future. Also, the conventional chemicals/fumigants being used world over are being restricted globally due to the environmental and health hazards posed by these toxic chemicals. Therefore, there is an urgent need to develop safe alternatives that have potential to replace toxic fumigants. In view of these, emphasis is on developing physical phytosanitary treatments as ecofriendly strategy for controlling the pests.

In the present era, efforts are being made for sustainable agriculture, improved nutrition and food security through peaceful uses of atomic energy. Irradiation as a phytosanitary measure has gained increasing acceptance in recent years. It has been recognized as a versatile treatment with broad spectrum activity against different pests. Significant contributions have been made worldwide using radiations for sanitary and phytosanitary purposes. The International Plant Protection Convention has also given acceptance to irradiation as a quarantine treatment in the year 2003 and has provided guidelines for the use of irradiation as a phytosanitary measure in International Standards on Phytosanitary Measures-18. Irradiation promises rapid, efficacious quarantine treatment for many pests with minimal damage to commodity quality. It appears to be a viable, non-chemical and residue-free strategy.

Gamma radiation is an approved and accepted technology for quarantine of the commodities for number of years but sometimes adversely affect the quality of the host commodities. Internationally, now new radiation generating sources as Electron beam are being explored to meet import standards of quality and quarantine. Low energy electrons with low penetration capacity have been reported to effectively disinfect grains, pulses, spices, dehydrated vegetables, tea leaves with considerably less quality deterioration than other disinfection techniques. The dose required for insect disinfestation is expected to be lower than that of disinfection. Disinfestation of grains and pulses with soft electrons potentially will have less deleterious effects on commodity quality than irradiation with gamma rays or high-energy electrons.

Electron Beam Technology

Electron beam (E-beam) technology has application in health care, agriculture and food preservation. Bhalla *et al.* (2009) has given a brief account of the electron beam technology and its use for agricultural purposes. The E-beam is a stream of high-energy electrons, propelled out of an electron gun. The beam, a concentrated highly charged stream of electrons, is generated by the acceleration and conversion of electricity. The E-beam has a machine source and can be simply switched on or off. No radioactivity is involved. E-beam irradiation uses electricity (not radioactive isotopes). E-beam radiation is a form of ionizing energy that is generally characterized by its low penetration and high dosage rates and is directional. The electrons are generated by equipment referred to as Accelerator, which is capable of producing beam that is either pulsed or continuous. E-beam comes from a filament, most commonly Tungsten hairpin gun.

Accelerators

E-beam accelerators are reliable and durable equipments and used in diverse industries to enhance the physical and chemical properties of the material and to reduce undesirable contaminants. Different end use areas need accelerators with different energies. Application of accelerator depends on its energy. Industrial electron accelerators can be classified as low-energy, medium-energy, and high-energy machines, based on the energies of the electrons produced. Electrons produced from accelerators with energies that are less than 1 MeV are classified as low-energy electrons. The low energy/soft electrons in the energy range of 60-200Kev is used for surface decontamination of seeds and spice. Medium-energy machines produce electrons with energies in the region 1 to 5 MeV, whereas high-energy accelerators produce electrons with energies that are greater than 5 MeV. A medium energy radiation in the range of 1-5MeV is used for grain disinfestations. Fruits, vegetables, fish, meat and fresh and frozen products require high radiation in the range of 5-10MeV. Use of E-beam of 10Mev is very efficient and inexpensive if the product presented to the beam has right density and thickness. The choice of irradiator depends on type of commodity, loose/packed, throughput required, thickness and shape of product, size of container, packaging density of the product, techno-economic feasibility.

Important advantages of electron beam technology include:

The technology has no deleterious environmental effects and is thus an environmentally friendly process.

☆ No load preconditioning needed

☆ No chemical/biological substances are involved so there are no undesirable chemical residues in the commodity

☆ E-beam irradiation uses electricity, not radioactive isotopes and therefore there is no radioactive waste

☆ Shorter exposure time facilitates fast processing

☆ No aeration process needed after sterilization

☆ Less expensive than other radiation techniques - in some cases

☆ Limited to site of treatment and period during which applied

Limitations of electron beam technology include:

☆ Dose uniformity; wide dose range may be observed, especially on high density loads

☆ Orientation of components is important

☆ Discoloration may occur on some materials

☆ High electric power consumption

☆ Complexity and potentially high maintenance

E-Beam Irradiation

Typical industrial applications involve the use of electrons with energies ranging from 3 to 10 MeV. The diameter of the electron beam which comes out from accelerator window is generally very less as compared to the product surface to be irradiated, hence the electron beam is scanned in one direction with a time varying magnetic field to achieve the wide radiation field, and the product is moved along the transverse direction with a suitable conveyor system. The radiation field profiles depend on electron beam size, energy/spectrum, beam current, beam scanning frequency, material of transmission foil and its thickness, air path etc. Dose uniformity on the product surface can be controlled with suitable combination of current profile in the scanning magnet and speed of conveyor system (Petwal *et al.*, 2004; Bhalla *et al.*, 2009).

As material passes beneath/in front of E-beam, energy from electrons is absorbed (Figures 12.1 and 12.2). The energy absorbed is referred as absorbed dose. This energy alters various chemical and biological bonds within material. Energy of E-beams is measured as amount of energy released by electron gun, in electron volts (eV). E-beam is less penetrating and the penetration depends upon the density of the material. Electrons spend their energy rapidly, create virtually no heat and dissipate leaving no residue. Speeding electrons damage/destroy microbial pathogens in and on food upon collision with DNA of organism. Measuring and monitoring of absorbed dosage is a critical point for application of irradiation process. The absorbed dose is measured using dosimeters.

In India, applications of radiation technology have been found in areas of health care, agriculture, food preservation, industry and environment. Both gamma and EM accelerator are being utilized for this purpose. Accelerators are available

Figure 12.1: Tray on the Conveyor Belt.

Figure 12.2: Irradiation of Samples.

at Raja Rammana Centre of Advanced Technology, Indore and Electron Beam Centre, Kahrghar, Navi Mumbai. For radiation processing of agricultural and food commodities, an MOU has been signed between Bhabha Atomic Research Centre and National Centre for Electron Beam Food Research and The Institute for Food Science and Engineering, Texas Agriculture Experimental Station, the Texas A and M University, Texas, USA for cooperation for advancement of Electron and X-ray irradiation technologies to promote food preservation, food safety and phytosanitary applications.

The new areas being explored include use of EB irradiation for surface treatment, radiation processing for curing of tyre components, radiation processed membranes for a variety of applications and radiation processing for treatment of environmental pollutants.

Efficacy of Electron Beam Irradiation as Phytosanitary Treatment against various Pests Infesting agricultural commodities

At low doses, irradiation could be used on a wide variety of foods to eliminate insect pests, as a replacement for fumigation with toxic chemicals that is routine for many foods now. It can also inhibit the growth of molds, inhibit sprouting, and prolong the shelf life. At higher doses, irradiation could be used on a variety of foods to eliminate parasites and bacteria that cause food borne diseases. A brief history of research on irradiation as a quarantine treatment is given by Burditt (1994). Important pests in grains and cereals include a wide variety of beetles, moths, weevils and others. The irradiation of at least one grain or cereal or processed grain product has been approved by 22 countries with rice being the most commonly approved cereal. Both gamma equipment-containing radioisotopes such as cobalt 60 and electron beam accelerators can be used to disinfest these commodities, depending on whether the product is available in bulk or bagged and other considerations etc. In case of bagged rice, research done by an Indonesian company under government supervision showed irradiation at 0.40 kGy minimum dose was sufficient to control the rice weevil. As a consequence, large quantities of rice have been irradiated commercially in Indonesia before entering government rice storage facilities. The irradiation of bagged grains or cereals is practical, effective and immediately available in many countries (http://www.food-irradiation.com/Grains.htm). For rice pests, 0.40-0.50 kGy can be used as the minimum dose for the pest control. The recommended maximum dose for rice is 2-3 kGy to prevent quality problems. Irradiation at doses used for disinfestation does not affect product quality of processed foods made from grains and cereals.

Extensive studies on the effect of different doses of EB on various pests (and stages thereof) *viz.*, *Tetranychus urticae* (Clas: Arachnida), *Thrips palmi* and *T. tabaci* (Order: Thysanoptera), *Spodoptera litura* (Order: Lepidoptera), green peach aphid, *Myzus persicae* (Order: Hemiptera), Comstock mealy bugs, *Pseudococcus comstocki* (Order: Homoptera), have been made by Dohino and co workers in Japan during 1993-1998. Gladon *et al.* (1997) has given an overview on the preservation of perishable products by using linear accelerator facility, Iowa State University, USA with approved irradiation dosage for the crops approved for radiation in the USA and worldwide. It includes the irradiation of strawberries and roses. Irradiation suppressed fungal growth (mainly *Botrytis cinerea*) on stored fruit, and doses of 1 and 2 kGy extended berry shelf life by 2 and 4 days, respectively. He concluded that electron beam irradiation has excellent potential for extending the shelf life of fresh strawberries by suppressing fungal growth, but cannot be used to eradicate or reduce fungal populations on roses without decreasing flower quality.

Hayashi *et al.* (2004) revealed that soft electrons at 60 keV effectively inactivated eggs, larvae and pupae of *Tribolium castaneum* and *Plodia interpunctella* and eggs of *Callosobruchus chinensis* at a dose of 1kGy. The adults of *T. castaneum* and *P. interpunctella* were inactivated by electron treatment at 5.0 kGy and 7.5 kGy respectively. Adults of *C. chinensis* survived at 7.5 kGy but were inactivated having lost the ability to walk at 2.5 kGy. Soft electrons at 60 keV could not completely inactivate the larvae of *C. chinensis* and small larvae (second instar) of *Sitophilus zeamais* inside beans and grains respectively because the electrons with low penetration did not reach larvae inside the host commodity. However, soft electrons at 60 keV inactivated eggs, larger larvae (IV instar) and pupae of *S. zeamais* in rice grains, which indicated that *S. zeamais* was exposed to electrons even inside the grains. The seeds of mungbean could be decontaminated with electrons at 75 keV. The results suggest that soft electrons necessary for decontamination did not adversely affect the sprouting of seeds or growth of sprouts. Also electrons at 60-70 keV reduced total microbial counts of alfalfa and radish sprouts to undetectable levels without any detrimental effect on germination ability. Low energy electrons eradicate microorganisms on the surface of seeds.

Though, the EB technology is well established as phytosanitary measure for agricultural commodities, its potential as a phytosanitary treatment for plant germplasm is to be exploited (*i.e.* the treatment should disinfest the material and at the same time should not affect the germination and viability of seeds). Soft electron treatment at 200 keV with 6 min of exposure time has been found to effectively control all life stages of *C. chinensis* with 80-100 per cent reduction in adult emergence. Even the weevils that survived the treatment were not able to multiply further. Hence, the treatment brings about total control of adzuki bean weevil. In addition, even the highest tested dose (20 kGy at 200 keV) of the soft electron treatment had no adverse effect on seed germination capacity and hence can be recommended even for treating the seed germplasm material and seed material to be used for sprouts (Reddy *et al.*, 2006). Electron beam irradiation at 500keV of seeds of 22 crops comprising cereals, legumes, oilseeds, vegetables and fibres revealed different effects on the seed quality parameters (percent germination and vigour index) of different crops (unpublished data). Further studies on different crops infested with their target pest/s and stages thereof revealed promising results.

Exploiting Potential of Electron Beam Irradiation as Phytosanitary Treatment against Insects in Seeds (Work done under the BRNS, DAE Project at ICAR-NBPGR)

Studies on the efficacy of electron beam irradiation as a phytosanitary treatment were conducted in some important crops against their major pests. Also the effect of electron beam on physiological and biochemical parameters of seed was studied. Irradiated seeds of two crops *viz.*, mungbean and cowpea were grown for observing chlorophyll mutations.

Efficacy of E-beam as Disinfestation of Mungbean Seeds Infested with *Callosobruchus maculatus*: A Case Study

The mungbean seeds artificially infested with different stages of *C. maculatus viz.*, egg, early larva, late larva, pupa and adult and uninfested seeds were irradiated at doses 170, 340 510, 680, 850, 1000, 1190 and 1360 Gy with electrons of 500 keV energy using a 750 keV, 20kW DC Accelerator developed by Raja Ramanna Centre for Advanced Technology, Indore, India. Irradiated infested seeds were tested for treatment efficacy based on the parameters of rate and number of adult emergence, their fecundity and emergence of F1 progeny for developmental stages of pest infesting seed; and days to mortality, fecundity of survivors and F1 progeny for adults infesting the seeds.

The uninfested irradiated/non-irradiated seeds were also tested for physiological parameters *viz.*, percent seed germination, shoot length, root length, seedling vigour (SV) and vigour index (VI) and biochemical parameters as electrical conductivity (EC) and tetrazolium test. Mungbean seeds irradiated at different doses were grown at NBPGR, New Delhi and observed for chlorophyll mutations.

Insects irradiated at 1190 and 1360 Gy at 500 keV showed 62.85 and 74 percent mortality, respectively, within 5 h after exposure to radiation in comparison to 0 per cent mortality in control. Complete mortality occurred within two days of irradiation at 1360Gy as compared to 13 days in control. Adult emergence from seeds infested with different stages was negligible and eggs laid by beetles that survived treatment (doses 340 Gy onwards) did not develop into adults. Irradiation with electrons at above doses had no significant effect on percent seed germination but affected the seedling vigour and vigour index. The lower doses *viz.*, 170, 340 and 510 Gy showed stimulatory effect on the seedling vigour and VI but higher doses affected the VI. The electrical conductivity of seed leachate and viability percentage based on tetrazolium test confirmed the results of germination test. Further no chlorophyll mutation was observed in the plants raised from the irradiated seeds. Results clearly demonstrate that low energy electron irradiation was an effective method for the management of *C. maculatus* adults. Sterility is caused at lower doses which had positive effect on seed quality parameters (Singh *et al.*, 2009; Bhalla *et al.*, 2010a; Bhalla *et al.*, 2010).

Similar results were also obtained in case of the soybean and chickpea seeds infested with *C. chinensis* The effective dose controlling all life stages/causing sterility in adults of *C. chinensis* had no adverse effect on seed germination, shoot and root vigour and vigour index (Singh *et al.*, 2014 and Bhalla *et al.*, unpublished data). Further studies on different crops infested with the target pest/s are under progress and are revealing promising results.

Ionizing radiation, either from isotope or machine generated sources, has been suggested as an alternative to methyl bromide and other chemical fumigants and is commercially used in small applications. These phytosanitary measures comprise the largest area of research in quarantine. The use of electrons from electron beam generators presents fewer health hazards than does the use of gamma rays, because of several advantages mentioned above. Taking into account these factors, soft electron radiation has a great potential for use in the disinfestations of food grains. Nevertheless, there are many avenues which merit further exploration. The adoption of particular treatment depends on the nature of pest and commodity to be treated, its level of effectiveness, convenience and ease of use and economic considerations. These measures need to be used in developing the disinfestation protocols to meet the international obligations under the WTO regime. EB is an emerging technology which is expected to have positive impacts on agriculture in terms of seed quality improvement/maintenance. It has the approval of the Agreement on Sanitary and Phytosanitary (SPS) under the World Trade Organization (WTO) and therefore can be adopted as an SPS measure in international trade.

References

Bhalla Shashi, Srinivasan K, Singh Subadas, Thakur Manju, Pramod R, Dwivedi J, Bapna S C, Sharma Arun and Sharma S K 2010a. Low energy electron irradiaton: a technology for quarantine disinfestation of green gram seeds against *Callosobruchus maculatus.* Presented in *3ʳᵈ Asia Pacific Symposium on Radiation Chemistry (APSRC-2010) and DAE BRNS 10ᵗʰ Biennial Trombay*

Symposium on Radiation and Photochemistry (TSRP-2010) September 14-17, 2010, Lonavala, India.

Bhalla Shashi, Srinivasan K, Singh Subadas, Thakur Manju, Pramod R, Dwivedi J, Bapna S C, Sharma Arun and Sharma S K 2010b. Low energy electron irradiaton: a novel technology to enhance quality of soybean seeds. Presented in *3rd Asia Pacific Symposium on Radiation Chemistry (APSRC-2010) and DAE BRNS 10th Biennial Trombay Symposium on Radiation and Photochemistry (TSRP-2010)* September 14-17, 2010, Lonavala, India.

Bhalla Shashi, Srinivasan K, Singh Subadas, Kumar Neeraj, Pramod R, Bapna S C, Dwivedi J, Sharma Arun and Khetarpal R K 2010. Effects of electron beam irradiation on germination and vigour of greengram seeds. *Seed Research* **38:** 16-20.

Bhalla Shashi, Khetarpal R K and Sharma S K 2009. Electron beam irradiation as phytosanitary measure for the safe movement of agricultural commodities, pp.12-126. *In:* Singh B, A Ananad and KR Koundal (eds.) *Training Manual of ICAR Winter School on Training and Capacity Building on Applications of Ionizing and Non ionizing Energies in Agriculture,* IARI, New Delhi 208 pp.

Bhalla Shashi, Sharma S K, Srinivasan Kalyani, Singh Subadas, Thakur Manju, Dwivedi J, Bapna S C and Khetarpal R K 2010a. Electron Beam: A novel technique to enhance quality of chickpea seeds. *In: Abstracts National Symposium on Post Harvest Management,* Karnal, Feb. 26-28, 2010, 50 pp.

Burditt Jr A K 1994. Irradiation, p.101-117. In: J L Sharp and G J Hallman (eds.), *Quarantine treatments for pests of food plants.* Boulder, CO. Westview Press, 290 pp.

Dohino T 1995. Effects of electron beam irradiation on Comstock mealybug, *Pseudococcus comstocki* (Kuwana) (Homoptera: Pseudococcidae). *Research Bulletin of the Plant Protection Service, Japan* **31:** 31-36.

Dohino T, Tanabe K, Masaki S and Hayashi T 1996. Effects of electron beam irradiation on *Thrips palmi* Karny and *Thrips tabaci* Lindeman (Thysanoptera: Thripidae). *Research Bulletin of the Plant Protection Service, Japan* **32:** 23-29.

Dohino T, Masaki S, Takano T and Hayashi T 1997. Effects of electron beam irradiation on sterility of Comstock mealybug, *Pseudococcus comstocki* (Kuwana) (Homoptera: Pseudococcidae). *Research Bulletin of the Plant Protection Service, Japan* **33:** 31-34.

Dohino T, Matsuoka S I, T T akano and Hayashi T 1998. Effects of electron beam irradiation on *Myzus persicae* (Sulzer) (Homoptera: Aphididae). *Research Bulletin of the Plant Protection Service, Japan* **34:** 15-22.

Gladon R J, Reitmeier C A, Gleason M L, Nonnecke G R, Agnew N H and Olsen D G 1997. Irradiation of horticultural crops at Iowa State University. *Horticultural Science* **32:** 582-585.

Hayashi T, Imamura T, Todoriki S S, Miyanoshita A and Nakakita H 2004. Soft-electron treatment as a phytosanitary measure for stored products pests. In:

Proceeding of final Research Coordination Meeting organized by FAO/IAEA in 2002. IAEA –Tecdoc-1427.

http://www.food-irradiation.com/Grains.htm Control of pests of grains, cereals and pulses by irradiation by Michelle Marcotte accessed on 11.9.2006.

Petwal V C, Pramod R, Kaul A, Bapna S C and Soni H C 2004. Electron beam dosimetry study for 750keV DC accelerator. *INSAC-04.*

Reddy P V R, Todoriki S, Miyanoshita A, Imamura T and Hayashi T 2006. Effect of soft electron treatment on adzuki bean weevil, *Callosobruchus chinensis* (L.) (Col., Bruchidae). *Journal of Applied Entomology* **130:** 393-399.

Singh Subadas, Bhalla Shashi, Bapna S C, Kumar Neeraj and Thakur Manju 2009. Efficacy of Electron Beam irradiation against pulse beetle, *Callosobruchus maculatus* (F.) (Coleoptera: Bruchidae). In: *Souvenir and Abstracts* (eds.) RK Tyagi, Bhag Mal, SK Sharma, RK Khetarpal, P Brahmi, N Singh, V Tyagi, S Archak, K Gupta and A Agarwal), Indian Society of Plant Genetic Resources, New Delhi, 370 pp.

Chapter 13

Geographical Information System (GIS) for Managing Plant Genetic Resources with Special Reference to Crop Health Management

N Sivaraj[1]*, S R Pandravada[1], V Kamala[1], N Sunil[2],
K Anitha[1], K Rameash[1], Babu Abraham[1], B Sarath Babu[1],
S K Chakrabarty[1], K S Varaprasad[3] and P C Agarwal[4]

[1]ICAR-National Bureau of Plant Genetic Resources,
Regional Station, Hyderabad – 500 030, Telangana State
[2]Winter Nursery Centre, ICAR-Indian Institute of Maize Research,
Hyderabad – 500 030, Telangana State
[3]ICAR-Indian Institute of Oilseeds Research,
Hyderabad – 500 030, Telangana State
[4]ICAR- National Bureau of Plant Genetic Resources,
New Delhi – 110 012
*E-mail: sivarajn@gmail.com

ABSTRACT

Management of Plant Genetic Resources (PGR) at national level involves enormous data and analysis of data is crucial to the effectiveness of its management process and can add significantly to the value this PGR. Geographical Information system (GIS) technology can be leveraged to meet the challenges and obtain results/output, which facilitate enhanced decision support in PGR management including planning. The sustainable management of plant genetic resources is of great concern as increasing population and rapid technological strides are putting enormous pressure on country's nutritional and food security. The major areas in which uses of GIS described are (i) inventorisation/ mapping, (ii) collection strategies, (iii) conservation strategies and (iv) crop expansion strategies in pest free region. GIS technology can be effectively used in planning field explorations for collecting agro-biodiversity, design and management of in-situ conservation sites, identify ecogeographical gaps in existing ex-situ germplasm collections, site identification for germplasm evaluation and regeneration; identifying geographic regions which are likely to contain specific desired traits, taxa or habitats of interest. Potential uses of GIS in PGR management and Crop health management have been highlighted and discussed in the chapter.

Keywords: *Plant genetic resources, Germplasm, GIS, Pest management.*

Plant genetic resources (PGR) are vital for crop improvement and ensuring food security. They comprise the diversity of genetic material contained in traditional cultivars, modern varieties as well as crop wild relatives and other economically important plant species that can be used as food, feed, fibre, clothing, shelter, wood, timber, energy etc. These natural resources represent both the raw material used in the production of new cultivars either through classical plant breeding or through biotechnology - and are a reservoir of genetic adaptability, which acts as a buffer against potentially harmful environmental and economic change. There are currently several underutilized plant species and varieties displaying traits of interest to meet present and future needs, while the value of many other plant species is yet to be discovered.

India has a rich and varied heritage of plant biodiversity, encompassing a wide spectrum of habitats ranging from tropical rain forests to alpine vegetation and from temperate forests to coastal wetlands (Gautam, 2004). The Indian sub-continent is one of the eight centres of origin (Vavilov, 1950), and is one of the 12 mega diversity centres of the world with 11.9 per cent of the world flora.India is endowed with rich PGR wealth of 49, 219 higher plant species including 5, 725 endemic species belonging to 141 genera under 47 families (Nayar, 1980). Of these endemic species, 3500 are found in the Himalayan region and 1600 in the Western Ghats (Arora, 1991).The Indian region is an important centre of origin and diversity of nearly 160 domesticated plant species of economic importance, more than 350 species and their wild relatives and over 9,500 species of ethno-botanical interest. It has about 30,000-50,000 landraces of rice, pigeonpea, mango, turmeric, ginger, sugarcane, etc. and ranks seventh in terms of contribution to world agriculture. An estimated 1000 wild edible plant species are exploited by native tribal communities. These include 145 species of roots and tubers, 521 of leafy vegetables/greens, 101 of buds and flowers, 647 of fruits and 118 of seeds and nuts (Arora and Pandey, 1996). Plant genetic resources of potential value are being lost at an alarming rate due to habitat destruction, land degradation, over exploitation of water resources, forestry practices, urban expansion, changing patterns of diversity, changing social and cultural norms, and adoption of improved varieties and technologies of intensive agriculture by the farmers (Gautam, 2004). A total of 4.0 lakhs of germplasm accessions are conserved in the seed bank *ex-situ,* located in National Gene Bank, New Delhi. Management of such huge PGR demands tools and techniques which could be effective and enhance decision support for utilisation and conservation.

The coming in force of the Convention on Biological Diversity (CBD), established in 1992, biodiversity in general including the wealth of plant genetic resources occurring within a country has become the property with sovereign rights established to the nation concerned changing the existing scenario of biodiversity as human heritage. Consequently, PGR conservation, sustainable use and equitable sharing of benefits arising out of such use have become the responsibility of sovereign nations. India is a signatory to CBD and International Treaty on Plant Genetic Resources for Food and Agriculture (ITPGRFA), with a list of 32 species of crop plants which are needed/essential for food and agricultural importance. Consequently, India confirmed its commitment towards sustainable PGR management for global food and fodder security. Thus, India has become a partner of the Global Plan of Action for the Conservation and Sustainable Utilization of Plant Genetic Resources for Food and Agriculture (GPA).

GIS technology helps researchers worldwide to assemble, store and retrieve large amounts of spatial data and other associated information related to Integrated Pest Management and other relevant approaches for managing plant health problems. Thus, the technology allows researchers to manipulate, analyse and display the spatial patterns of variables (environment, economy, socio-cultural aspects etc.), which are having direct and indirect influence in solving problems of crop health management.

What is Geospatial Technology?

An organized collection of computer hardware, software, geographic data, and personnel designed to efficiently capture, store, update, manipulate, analyze, and display geographically referenced information.

Possible Uses of GIS in Plant Health Management

Geographic information systems have the ability to display layers of spatially referenced information on various pests. One of the most useful applications in plant disease management work is the display of point data as a layer over other layers showing background information or attribute data for developing strategies for sustainable plant health management. The GIS tools offer opportunities for cost-effective and efficient targeting of oppress interventions so as to make effective plant protection strategies. In monitoring, GIS can be used to determine the spatial extent of a disease, to identify spatial patterns of the disease and to link the disease to auxiliary spatial data, *viz.*, weather, vegetation, ecogeographic conditions.

Pest Monitoring and Detection

☆ Data visualization and query

☆ Survey data collection, management, and analysis

☆ Risk and pathway analysis

  ☐ What area has highest risk for a pest introduction?

  ☐ Prediction models can be generated

☆ Change detection in case of Invasive Alien Species and others

Why Should we use GIS in Plant Health Management?

☆ Economic benefit

☆ Proactive approach in safeguarding Indian Agriculture

☆ Quality control assistance

☆ Decision support system

Possible Outputs of GIS Use in Plant Health and PGR Management

☆ Update passport information for pest/PGR collecting sites with respect to geo-reference information.

☆ Map the pest range/crop diversity collected for individual countries and also on global basis.

☆ Analyze crop/pest diversity collected for different passport and characterization information.

☆ Complementary diversity analysis for combination of traits.

☆ Based on past collecting information, identify potential matching sites for cultivation/pest spread

☆ Classify collections based on climatic adaptation.

☆ Provide climatic information (Monthly rainfall, minimum and maximum temperature) for individual collecting sites.

☆ Providing climate maps for various climatic parameters and their combinations as well as for altitude of collecting sites.

☆ Providing guidelines to further develop collecting strategies for new collections as well as for re-collecting of germplasm and to study/identify pest free zones.

Some Examples

☆ Adult grasshopper hazard analysis (in USA) – identifies patterns in grasshopper survey that may predict future population increases – Understand patterns in grasshopper survey as related to environmental conditions (*i.e.* climate, soil, vegetation)

☆ Area-wide insect pest management (Klassen, 2000; Beckler *et al.*, 2005; Huang *et al.*, 2008)

☆ Asian Gypsy moth trapping model (in USA) - GIS to identify those areas most likely to have AGM activity – to improve and/or validate existing AGM trapping locations

☆ European Corn Borer (ECB) and Corn Earworm (CEW) management in vegetable crops (New Jersey, USA) (Kristian *et al.*, 2001)

☆ Insect pest management in Tea (Assam, India) (Hazarika *et al.*, 2009)

☆ Lepidopteran cereal stem borer management in Africa (Rami *et al.*, 2002)

☆ Mapping of pest distribution data (from pest interceptions of import quarantine)

☆ Screening for anthracnose disease resistance in horse gram (India) (Udaya Sankar *et al.*, 2015)

☆ Sugarcane wooly aphid spread in south India (Ganeshaiah *et al.*, 2003)

☆ Spruce bark beetle management in forest lands of Denmark (Wichmann and Ravn, 2001)

☆ Spatial distribution of seed borne fungi on *Pongamia pinnata* (India) (Anitha *et al.*, 2010)

☆ To quantify area change in Invasive alien species spread

☆ Western tarnished plant bug management in cotton (in Arizona, USA) (Carrière *et al.*, 2006)

Geographic information system (GIS) integrates hardware, software, and data for capturing, managing, analyzing, and displaying all forms of geographically referenced information. GIS allows us to view, understand, question, interpret, and visualize data in many ways that reveal relationships, patterns, and trends in the form of maps, globes, reports, and charts (*www.esri.com*). Thus, GIS is a database management system that can simultaneously handle data representing spatial objects and their attribute data. GIS can be effectively used in Plant health and PGR management particularly in the areas namely a) inventorisation/mapping, b) pest free collection strategies, c) pest free conservation strategies and d) crop expansion strategies.

Inventorisation/Mapping Strategies

Priority Action 1 of GPA calls for increased surveying and inventorying of plant genetic resources for food and agriculture. Datasets of PGR with pest attributes, identity and geo-reference data of relevant point locations are prerequisites for GIS mapping. GIS can be effectively used in preparing pest distribution maps of interest. Some of the GIS Software used for plant health management is provided in Table 13.1.

GIS mapping has been successfully used in assessing biodiversity and in identifying areas of high diversity in *Phaseolus* bean (Jones *et al.*, 1997); coconut (Bourdeix *et al.*, 2005); wild potatoes (Hijmans and Spooner, 2001); wild Arachis (Jarvis *et al.*, 2003); horsegram (Sunil *et al.*, 2008); *Jatropha curcas* (Sunil *et al.*, 2009); linseed (Sivaraj *et al.*, 2009; 2012); rapeseed-mustard (Semwal *et al.*, 2012); safflower (Dikshit *et al.*, 2012, Sivaraj *et al.*, 2012); sesame (Spandana *et al.*, 2012), blackgram (Babu Abraham *et al.*, 2010); piper (Parthasarathy *et al.*, 2008); *Canavalia* fatty acids (Sivaraj *et al.*, 2010); medicinal plants (Varaprasad *et al.*, 2007); wild species (Greene *et al.*, 1999); plant genetic resources collection (Greene and Hart, 1996, Hart *et al.*, 1996); *Spondias* (Miller *et al.*, 2006) and agrobiodiversity in general for South East coastal India (Varaprasad *et al.*, 2008). Similarly, GIS has been successfully used in pest management by several researchers. Area-wide insect pest management (Klassen, 2000; Beckler *et al.*, 2005; Huang *et al.*, 2008); European corn borer and corn earworm (Kristian *et al.*, 2001); Tea insect pests (Hazarika *et al.*, 2009); cereal stem borer (Rami *et al.*, 2002); mapping anthracnose disease in horse gram (Udaya Sankar *et al.*, 2015); sugarcane wooly aphid (Ganeshaiah *et al.*, 2003); spruce beetle (Wichmann and Ravn, 2001); seed borne fungi (Anitha *et al.*, 2010); plant bug (Carrière *et al.*, 2006).

Table 13.1: GIS Software Used for Sustainable Plant Health Management

Name of GIS Software	Function
Arc GIS	Tools for building applications (including genetic resources management)
Biomapper	GIS tool kit to model ecological niche and habitat suitability
CLIMEX	Assess the risk of pest establishing in a new location
Degree	Spatial data infrastructures
GARP	Predict and analyse species distribution
DIVA-GIS	Maps of species distribution and analysis
DSSAT	Software that combines crop, soil and weather data bases into standard formats for access by crop models and application programs
EcoSim	Null model software for ecologists
EstimateS	Statistical estimation of species richness and shared species from samples
FloraMap	Likely distribution of wild species in nature
GBIF MAPA	Species Richness Assessment (SRA)
Geo Da	Spatial data analysis
GPSphotolinker	Save location and GPS position data to a photo
gvSIG	Manage geographic information
HyperNiche	Multiplicative habitat modelling
Maxent	Species and habitat modeling
Quantum GIS	Spatial data analysis
S-Distance	Spatial decision support system
SAM (Spatial Analysis Macroecology)	Statistical tools for spatial analysis
SPADE (Species prediction and Diversity)	Spatial data analysis

Pest Free Germplasm Collection Strategies

Germplasm exploration and pest free collections are planned generally based on available databases of passport information. Passport information includes an identity to the collection, specific location of collection, details of habit/habitat and other reference data. GIS can be effectively used in preparing distribution maps of species, probable location of the collection sites, gap analysis, analyzing diversity rich pockets etc. GIS can be used to link the passport database with district and state map layers to analyse what has been explored and collected and from where and what are the gaps in terms of areas to be explored and germplasm needs to be collected. Thus, to plan future exploration programmes which are trait specific/region specific GIS can be used effectively. Mapping the spatial distribution of disease free target species along with the prevailing knowledge systems of communities can be effectively carried out using GIS. Traditionally, tribals/farmers are the custodians of plant genetic resources and have also developed huge knowledge systems over years. These traditional knowledge systems form an integral part of the all the communities and more specifically tribal communities. Indigenous traditional knowledge associated with the plant genetic resources, their time of cultivation, system of cultivation abd its relation to the environment form a vital part of the tribal system. Such systems which co-evolved with the nature provide the food and nutritional security of the tribal communities. Documentation followed by validation of such crop genetic resources related ITK would make available such secure sources of ethnic systems to be harnessed for benefit of all. GIS and other specialized computer program (*e.g.* FloraMap) along with associated data can be used to map the predicted distribution of plant/insect species or areas of possible climatic adaptation of organisms in the wild (Jones *et al.*, 2002). Also, GIS can play an important role in the management of large and complex PGR datasets (Guarino *et al.*, 2001). Guarino (1995) discussed the use of GIS to develop strategies for collecting germplasm. For example, collection regions can be mapped to identify areas with desired ecogeographic attributes such as acid soils or climate extremes (Hart *et al.*, 1996).

Pest Free Conservation Strategies

Complementary pest free conservation strategies include the protection of wild species, plant populations and traditional crop varieties where they have evolved (*in situ* conservation), with the collection and preservation of inter- and intra specific diversity in gene banks and botanical gardens (*ex situ* conservation). *Ex situ* genetic resource collections maintain germplasm in the form of seed or live plants, representing current, obsolete and primitive crop varieties, wild and weedy relatives of crop species, and wild species collected or augmented from around the world. GIS can be effectively used for genetic resources conservation in the following areas:

☆ Identifying gaps for conservation in both *ex-situ* and *in-situ*

☆ Design and management of *on-farm in-situ* conservation sites

Geographical information systems (GIS), climate change models and geographical distribution data of crop plants and their wild relatives may be used to predict the impact of a changing climate on plant genetic resources (PGR), conservation and use, in various agri-horticultural crops important in Indian crop improvement programmes.

Crop Expansion Strategies in Pest Free Regions

GIS can play a crucial role by way of managing large data sets, identifying suitable locations for multiplication and evaluation of germplasm introduced from other countries. Morphological descriptors/genetic variations may be linked with environmental attributes using GIS for selecting potentially useful germplasm accessions for crop expansion (Pederson *et al.*, 1996). Exotic germplasm of several agri-horticulutral crops (Tef, Kiwi, Olive etc.) can be introduced into suitable agro-climatic regions of our country after assessment using GIS technologies *viz.* Flora map, DIVA-GIS, GARP, MaxEnt, Ecocrop. Similarly, suitable regions for cultivation of economically important exotic crops in pest free regions could be assessed using GIS.

Conclusion

Geographic information systems (GIS) provide valuable tools in monitoring, predicting, managing and combating the spread of insect pests and diseases for sustainable plant health management.GIS can also be used to predict the projected spread of diseases, to provide input for risk assessment models in pest control and in quantifying changing thresholds of pests and diseases due to climate change. Thus, Geographical information system can be effectively used for plant health management and plant genetic resources management with enhanced output/ deliverables for effective decision support and planning.

References

Abraham B, Kamala V, Sivaraj N, Sunil N, Pandravada S R, Vanaja M and Varaprasad K S 2010. DIVA-GIS approaches for diversity assessment of pod characteristics in black gram (*Vigna mungo* L. Hepper). *Current Science* **98:** 616-619.

Anitha K Sunil N, Sivaraj N, Chakrabarty S K, Babu Abraham, Viond Kumar, Suresh Kumar G, Varaprasad K S and Khetarpal R K 2010. Spatial distribution of seedborne fungi on *Pongamia pinnata*: A DIVA-GIS analysis with reference to *Macrophomina phaseolina*. *Indian Journal of Plant Protection* **38:** 67-72.

Arora R K 1991. Plant diversity in Indian gene centre. *In:* Plant genetic resources-conservation and management (eds. Paroda R S and Arora R K). IPGRI, Regional office for South Asia, New Delhi. pp. 25-54.

Arora R K and Pandey A 1996. Wild edible plants of India: Diversity, conservation and use. National Bureau of Plant Genetic Resources, New Delhi, India. 294 pp.

Beckler A A, French B W and Chandler L D 2005. Using GIS in Areawide Pest Management: A Case Study in South Dakota. *Transactions in GIS* **9:** 109-127.

Bonman J M, Bockelman H E, Jin Y, Hijmans R J and GironellaA I N 2007. Geographic distribution of stem rust resistance in wheat landraces. *Crop Science* **47:** 1955-1963.

Bourdeix R, Guarino L, Mathur P N and Baudouin L 2005. Mapping of coconut genetic diversity. *In:* Coconut Genetic Resources (eds. Pons Batugal, Ramanatha Rao V and Jeffrey Oliver). IPGRI. pp. 32-43.

Carrière Y, Ellsworth P C, Dutilleul P, Ellers-Kirk C, Barkley V and Antilla L 2006. A GIS-based approach for area wide pest management: The scales of *Lygushesperus* movements to cotton from alfalfa, weeds, and cotton. *Entomologia Experimentaliset Applicata*118: 203-210.

Dikshit N, Abdul Nizar M and Sivaraj N 2012. Evaluation and diversity analysis of safflower germplasm in relation to morpho-agronomic characteristics. *Journal of Oilseeds Research* **29 (Special Issue):** 17-23.

Ganeshaiah K N, Barve N, Nilima N, Chandrashekara K, Swamy M and Uma Shanker R 2003. Predicting the potential geographical distribution of the sugarcane wooly aphid using GARP and DIVA-GIS. *Current Science* **85:** 1526-1528.

Gautam P L2004. Trends in plant genetic resource management. *In*: Plant genetic resource management (eds. Dhillon B S, Tyagi R K and Arjun Lal). Narosa Publishing House, New Delhi. pp. 18-30.

Greene S L and Hart T 1996. Plant genetic resource collection: an opportunity for the evolution of global data sets. http://www.ncgia.ucsb.edu/conf/SANTA_FE_CD-ROM/sf_papers/Greene_stephanie/sgreene.html Accessed 02.12.2014.

Greene S, Hart T and Afonin A 1999. Using geographic information to acquire wild crop germplasm: II. Post collection analysis. *Crop Science* **39:** 843-849.

Guarino L1995. Geographic information systems and remote sensing for the plant germplasm collector. *In*: Collecting Plant Genetic Diversity. Technical Guidelines (eds. Guarino L., Ramanatha Rao, V and Reid R). CAB International, Wallingford, UK, pp. 315–328.

Guarino L, Jarvis A, Hijmans R J and Maxted N 2002. Geographic information systems (GIS) and the conservation and use of plant genetic resources. *In*: Managing Plant Genetic Diversity (eds. Engels J M M, Rao V R, Brown A H D, Jackson M T). IPGRI, Rome, Italy. pp. 387-404.

Guarino L, Maxted N and Sawkins M 1999. Analysis of geo-referenced data and the conservation and use of plantgenetic resources. *In*: Linking Genetic Resources and Geography: Emerging Strategies for Conserving and Using Crop Biodiversity (eds. Greene S L Guarino L). CSSA Special Publication No. 27. ASA and CSSA, Madison,Wisconsin, pp. 1-24.

Hart T S, Greene S L and Afonin A 1996. Mapping for Germplasm Collections: Site Selection and Attribution. *Proceedings of the third international conference on integrating GIS and environmental modeling*. NCGIA, Santa Barbara, CA.

Hazarika L K, Bhuyan M and Hazarika B N 2009. Insect pests of tea and their management. *Annual Review of Entomology* **54:** 267-284.

Hijmans R J and Spooner D M 2001. Geographic distribution of wild potato species. *American Journal of Botany* **88:** 2101-2112.

Hijmans R J, Guarino L, Cruz M and Rojas E 2001. Computer tools for spatial analysis of plant genetic resources data 1. DIVA-GIS. *Plant Genetic Resources Newsletter* **127:** 15-19.

Huang Y, Lan Y, Westbrook J K, Wesley C and Hoffmann W C 2008. Remote Sensing and GIS Applications for Precision Area-Wide Pest Management: Implications for Homeland Security. Geospatial technologies and homeland security. *The Geo Journal Library* **94:** 241-255.

Jarvis A, Ferguson M E, Williams D E., Guarino L, Jones P G, Stalker H T, Valls J F M, Pittman R N, Simpson C E, and Bramel P 2003. Biogeography of wild

Arachis: assessing conservation status and setting future priorities. *Crop Science* **43**: 1100-1108.

Jones P G, Beebe S E, Tohme J and Galway N W 1997. The use of geographical information systems in biodiversity exploration and conservation. *Biodiversity and Conservation* **6**: 947-958.

Jones P G and Gladkov A 1999. Flora Map: A Computer Tool for the Distribution of Plants and Other Organisms in the Wild. CIAT, Cali, Colombia.

Jones P G, Guarino L and Jarvis A 2002. Computer tools for spatial analysis of plant genetic resources data: 2. Flora Map. *PGR Newsletter* **130**: 1-6.

Klassen W2000. Area-wide approaches to insect pest interventions: history and lessons. Pp. 21-38 in Teng-Hong Tan (ed.), Joint Proceedings of the FAO/IAEA International Conference on Area-Wide Control of Insect Pests, May 28-June 2, 1998, and the Fifth International Symposium on Fruit Flies of Economic Importance, June 1-5, 1998. I.A.E.A., Penerbit UniversitiSains Malaysia, Pulau, Pinang, Malaysia, 782 pp.

Kristian E Holmstrom,Marilyn GHughes, Sarah DWalker, Wesley L Kline and Joseph Ingerson-Mahar 2001. Spatial Mapping of Adult Corn Earworm and European Corn borer populations in New Jersey. *Horticulture Technology* **11**: 103-109.

Miller Allison J and Knouft J H 2006. GIS-based characterization of the geographic distributions of wild and cultivated populations of the Mesoamerican fruit tree *Spondias purpurea* (Anacardiaceae). *American Journal of Botany* **93**: 1757-1767.

Nayar M P 1980. Endemism and pattern of distribution of endemic genera (angiosperm) in India. *Journal of Economic and Taxonomic Botany* **1**: 99-110.

Parthasarathy U, George J, Saji K V, Srinivasan V, Madan M S, Mathur P N and Parthasarathy V A 2008. Spatial analysis for Piper species distribution in India. *Plant Genetic Resources Newsletter* **147**: 1-5.

Pederson G A, Fairbrother T E and Greene S L 1996. Cyanogenesis and climate relationships in U.S. white clover germplasm collection and core subset. *Crop Science* **36**: 427-433.

Rami K, Overholt W A, Khan Z R and Polaszek A 2002. Biology and management of economically important Lepidopteran cereal stem borers in Africa. *Annual Review of Entomology* **47**: 701-731.

Semwal D P, Bhandari D C, Bhatt K C and Ranbir Singh 2012. Diversity distribution pattern in collected germplasm of Rapeseed-Mustard using GIS in India. *Indian Journal of Plant Genetic Resources* **26**: 76-81.

Sivaraj N, Sunil N, Pandravada S R, Kamala V, Vinod Kumar, Rao B V S K, Prasad R B N and Varaprasad K S 2009. DIVA-GIS approaches for diversity assessment of fatty acid composition in linseed (*Linum usitatissimum* L.) germplasm collections from peninsular India. *Journal of Oilseeds Research* **26**: 13-15.

Sivaraj N, Sunil N, Pandravada S R, Kamala V, Rao B V S K, Prasad R B N, Nayar E R, Joseph John K, Abraham Z and Varaprasad K S 2010. Fatty acid composition in seeds of Jack bean [*Canavalia ensiformis* (L.) DC] and Sword bean [*Canavalia gladiata* Jacq.)DC] germplasm from South India: A DIVA-GIS analysis. *Seed Technology* 32: 46-53.

Sivaraj N, Sunil N, Pandravada S R, Kamala V, Vinod Kumar, Babu Abraham, Rao, B V S K, Prasad R B N and Varaprasad K S 2012. Variability in linseed (*Linum usitatissimum*) germplasm collections from peninsular India with special reference to seed traits and fatty acid composition. *Indian Journal of Agricultural Sciences* 82: 102-105.

Sivaraj N, Pandravada S R, Dikshit N, Abdul Nizar M, Kamala V, Sunil N, Chakrabarty S K, Mukta N and Varaprasad K S 2012. Geographical Information System (GIS) approach for sustainable management of Safflower (*Carthamus tinctorius* L.) genetic resources in India. *Journal of Oilseeds Research* **29 (Special Issue):** 45 - 49.

Spandana B, Sivaraj N, John Prasanna Rao G, Anuradha G, Sivaramakrishnan S and Farzna Jabeen 2012. Diversity analysis of sesame germplasm using DIVA-GIS. *Journal of Spices and Aromatic Crops* **21:** 145-150.

Sunil N, Sivaraj N, Anitha K, Babu Abraham, Vinod Kumar, Sudhir E, Vanaja M and Varaprasad K S 2009. Analysis of diversity and distribution of *Jatropha curcas* L. germplasm using Geographic Information System (DIVA-GIS). *Genetic Resources and Crop Evolution* **56:** 115-119.

Sunil N, Sivaraj N, Pandravada S R, Kamala V, Raghuram Reddy P and Varaprasad K S 2008. Genetic and geographical divergence in horsegramgermplasm from Andhra Pradesh, India. *Plant Genetic Resources: Characterization and Utilization* **7:** 84-87.

Udaya Sankar A, Anitha K, Sivaraj, N, MeenaKumari K V S, Sunil N and Chakrabarty S K 2015. Screening of horsegram germplasm from Andhra Pradesh against anthracnose. *Legume Research*: Print ISSN 0250-5371. Online ISSN 0976-0571.

Varaprasad K S, Sivaraj N, Mohd Ismail and Pareek S K 2007. GIS mapping of selected medicinal plants diversity in the Southeast Coastal Zone for effective collection and conservation. *In*: Advances in Medicinal Plants (eds. Janardhan Reddy K, Bir Bahadur, Bhadraiah B and Rao M L N). Universities Press (India) Private Ltd. New Delhi. pp. 69-78.

Varaprasad K S, Sivaraj N, Pandravada S R, Kamala V and Sunil N 2008. GIS mapping of Agrobiodiversity in Andhra Pradesh. *Proceedings of Andhra Pradesh Akademi of Sciences.* Special Issue on Plant wealth of Andhra Pradesh. pp. 24-33.

Vavilov N I 1951. The Origin, Variation, Immunity and Breeding of Cultivated Plants. Ronald Press Company, New York.

Wichmann L and Ravn H P 2001. The spread of *Ips typographus* (L.) (Coleoptera, Scolytidae) attacks following heavy windthrow in Denmark, analysed using GIS. *Forest Ecology and Management* **148:** 31-39.

Chapter 14

Entomopathogenic Nematodes as a Component in Rice Pest Management: A Journey from Lab to Land

J S Prasad, A P Padmakumari*,
N Somasekhar and G Katti

*ICAR-Indian Institute of Rice Research, Rajendranagar,
Hyderabad – 500 030, Telangana
E-mail: padmakumariento@gmail.com

ABSTRACT

Studies were carried out on four indigenous entomopathogenic nematodes (EPN) in both laboratory and field conditions. Our studies indicated that the infective juveniles (IJs) of indigenous EPN when applied to soil could move from base to top of the plant and also infect the host insect, yellow stem borer (YSB) present within the rice stem. When exposed to UV rays for different time intervals, variation was observed in the mortality rates and infectivity of EPN to Galleria mellonella. The EPN species varied in their sensitivity to insecticides but at a concentration of 0.025 per cent of the tested insecticides viz., chlorpyrifos, and imidacloprid the mortality was less than 20 per cent. All the species were highly sensitive to cartaphydrohloride. Increase in the concentration of the insecticide resulted in increased mortality of Ijs. Sensitivity of EPN to adjuvants and insecticides varied with species and time of exposure. Field trials in microplots with three EPN species indicated that the spray of EPN reduced the white ear damage in rice by 23.9 to 37.7 per cent as compared to control with increase in grain yield. Our studies prove that entomopathogenic nematodes could be utilized as a component in rice IPM for the management of yellow stem borer.

Keywords: Indigenous entomopathogenic nematodes, Yellow stem borer, Rice, Field efficacy.

Introduction

Rice is a staple food crop inflicted by many biotic and abiotic stresses. Of the many biotic stresses, insect pests cause 25 per cent yield loss in rice amounting to Rs 240138 million rupees (Dhaliwal *et al.*, 2010). At present, timely application of chemical insecticides is the only viable option available for the management of two major lepidopteran pests, stem borers and leaffolder. Efforts towards development of alternative methods led to the identification of indigenous entomopathogenic nematodes (EPN) as potential bio control agents for the management of rice pests. Utilisation of EPN has raised intense interest globally mainly because of its potential, efficiency, exemption from registration and other impressive attributes conducive for easy use in rice ecosystem. Nematodes contribute to biological pest suppression through the actions of endemic populations (Campbell *et al.*, 1998; Duncan *et al.*, 2003) purposefully released using classical, inoculative, or inundative approaches (Parkman *et al.*, 1993; Grewal and Georgis, 1999, Shapiro-Ilan *et al.*, 2002). Efficacy of four species of indigenous entomopathogenic nematodes, *Metarhabditis amsactae* isolate Drr - Ma 1, *M. amsactae* isolate Drr –Ma2, *Steinernema asiaticum* and *Heterorhabditis indica* against the egg mass, larvae and pupae of yellow stem borer (YSB) both in laboratory and field conditions and against leaffolder in both laboratory and greenhouse (Prasad *et al.*, 2003; Katti *et al.*, 2003; Padmakumari *et al.*, 2005, 2007, 2008 and Sankar *et al.*, 2009) has been documented. The storability of these nematodes was reported by Gururaj Katti *et al.* (2006). Before taking up further field evaluations, it was felt pertinent to study the sensitivity of EPNs to UV light and insecticides, while to increase their stability for application as field sprays, the effect of adjuvants needed to be examined. The present paper reports the effect of insecticides, UV light and adjuvants on EPN survival, and also the performance of indigenous EPNs when applied in the greenhouse and field for reducing yellow stem borer damage.

Materials and Methods

Prior to utilization of EPN in field, the indigenous EPNs were suitably characterized for their efficacy against rice pests. The laboratory cultures of four species of EPN, *Metarhabditis amsactae* -Drr Ma1 (Earlier *Rhabditis* (*Oscheius*) sp), *M. amsactae*-Drr Ma2 (formerly *S. thermophilum*), *S. asiaticum* and *H. indica* were maintained on *Galleria mellonella* for use in further studies. Experiments were carried out under laboratory conditions to test the efficacy of these EPNs when exposed to UV light, adjuvants and insecticides before limited field trials were taken up.

Studies on Searching Ability and Mobility of EPNs

Experiments were carried out on 20 day old rice seedlings raised in pots. EPN of each species (100 nos) were inoculated at the base of the plant and the plants were covered with the mylar film. After 24 h, the water droplets and top foliage were observed for EPN under microscope. In another experiment, 20 infested stubbles were uprooted from the field at harvest through stratified random sampling and were brought to the laboratory. In each 500 ml beaker, four stubbles (25-30 tillers) with roots were placed and sufficient moisture was maintained. The stubbles were sprayed with 50,000 IJs of EPN. Five replications were maintained for each set up. After seven days, the stubbles were dissected out.

Effect of Insecticides on EPN

In view of the importance of insecticides in integrated pest management (IPM), a study was taken up under laboratory conditions to ascertain the compatibility of EPN with four selected insecticides *viz.*, chlorpyriphos, imidacloprid, cartap hydrochloride and monocrotophos, which are widely used in rice ecosystem. IJs of each EPN species (100 IJs/replication/treatment) were incubated in five concentrations (0.025 per cent, 0.05 per cent, 0.075 per cent, 0.1 per cent and 0.125 per cent) of test insecticides replicated thrice each. The surviving IJs were tested for their infectivity to *G. mellonella* to assess the effect of insecticide on EPNs.

Effect of UV Light on EPN

Experiments were carried out on the survival of IJs of *M. amsactae* -Drr Ma1, *S. asiaticum* and *H. indica* after exposure to UV light (Philipps TUV 30W/G30T8) for different periods *viz.*, 10 to 160 minutes at 10 min interval. Each time period was replicated thrice and 100 IJs per replication/EPN species were exposed for each time period. After exposure to specific time intervals, the mortality of IJs was assessed and the surviving IJs were tested for their infectivity to late instar *G. mellonella* larvae.

Effect of Adjuvants on Survival of EPN

To increase the field efficacy of the EPN, experiments were conducted with six adjuvants *viz.*, Labklin®, labolene®, Tween 20, Triton X 100, glycerol and liquid paraffin at three concentrations each *viz.*, 0.5 per cent, 1 per cent and 2 per cent. Hundred IJS of each EPN sp were incubated with the solutions at different time intervals of 1h, 2h, 3h and 4h and each treatment was replicated three times. Observations were recorded on the number of surviving IJs to estimate the per cent mortality. The surviving IJs were released on to late instar *G. mellonella* to test their infectivity.

Greenhouse Evaluation of EPN against YSB

Twelve potted plants were raised with five 25 days old seedlings of cultivar TN1 in each pot. Two five days old egg masses of YSB were pinned on to the leaf laminae nearer to the axil. Suspensions of *M. amsactae* Drr-Ma1, Drr- Ma 2 and *S. asiaticum* consisting of 2000 IJs/100 ml water with 0.1 per cent glycerol were sprayed on to the potted plants. Healthy potted plants with and without egg masses were maintained as controls for each treatment to assess the damage.

Evaluation of the Efficacy of EPN in Field

Field trials were carried out to prove the efficacy of EPN against yellow stem borer for two seasons (*Kharif* 2007 and *Rabi* 2007-08) on different varieties. Efficacy of Drr- Ma1, *H. indica* and Drr-Ma 2 against YSB was evaluated in microplots (1X1 m²) in the main field along with an insecticide cartap hydrochloride and water spray. In each of the seasons, nine micro plots were demarcated for each of the EPN species and YSB egg masses were pinned at booting stage of the crop to supplement natural infestation. The treatments were randomized and replicated nine times. Each micro plot was sprayed in the evening with EPN spray fluid @ one lakh IJs/m². The number of white ears (WE) in each microplot was counted to work out the percentage white ear damage and data were subjected to ANOVA. The pinned egg mass were examined after 10 days to check for hatchability and also recovery of nematodes.

Results and Discussion

Studies on Searching Ability and Mobility of EPNs

The results revealed that more than 70 per cent of the IJs of EPN applied to soil in the pots could be collected from the foliage of the potted rice plant within 24 hours of application. Similarly, EPNs inoculated on to rice stubbles could infect the YSB larvae inside the paddy stubbles. A week after application, EPN could be recovered from the infected larvae. One per cent of the larvae were infected but the lumen had many EPN. Hence, it was evident that the EPNs could not only migrate from soil to the plant foliage but were also able to locate the YSB larvae in the lumen of the stubble and infect successfully. The searching ability of the EPN plays a significant role in the management of cryptic pests like stem borer and leaf folder, which are always within the plant parts. Earlier studies also proved that the Drr- Ma1 and Drr- Ma 2 could successfully penetrate through the tuft of hairs covering the egg mass of yellow stem borer and infect the eggs (Katti *et al.,* 2003).

Effect of Insecticides on EPN

The tolerance of EPNs to test insecticides varied from species to species. The three EPN isolates tested were sensitive to all the test insecticides at doses higher than 0.025 per cent. The effect of insecticides on EPN IJs is depicted in Figures 14.1a-d. Among the two organophosphate insecticides that were tested, the IJs were more sensitive to monocrotophos as there was 100 per cent mortality within 8h of exposure as compared to chlorpyriphos where 100 per cent mortality was recorded 120h after exposure. Cartap hydrochloride caused toxicity to nematode IJs within 10 to 60 minutes after exposure. IJs survived upto 72 h of exposure in imidacloprid treatment. However, Devi Gitanjali (2011) reported that 24 h of exposure of *H. indica* IJs to monocrotophos was safe.

All the EPNs were sensitive to cartap hydrochloride a synthetic derivative of nereistoxin. Among the EPNs, Drr-Ma2 was more sensitive to cartaphydrochloride, monocrotophos and chlorpyriphos compared to *S. asiaticum* and *H. indica.* Based on IJs survival, *S. asiaticum* was highly sensitive to cartap hydrochloride, monocrotophos and imidacloprid. The low mortality of IJs indicates that it was compatible with chlorpyrifos at 0.025 per cent and 0.05 per cent. Insecticides affected the survival, infectivity as well as recovery of the IJs from the hosts. Our studies indicate that IJs incubated with monocrotophos could not infect *G. mellonella* larvae. Some of the IJs of Drr- Ma2 and *S. asiaticum* had taken longer time (up to 94.2 h and 106 h, respectively) to infect while some of them lost infectivity when exposed to cartap hydrochloride. IJs of *H. indica* were infective only at low concentration (upto 0.05 per cent). *Drr -Ma2* was sensitive to monocrotophos. *S. asiaticum* and *H.*

Figures 14.1a-d: Effect of insecticides on EPN Survival.

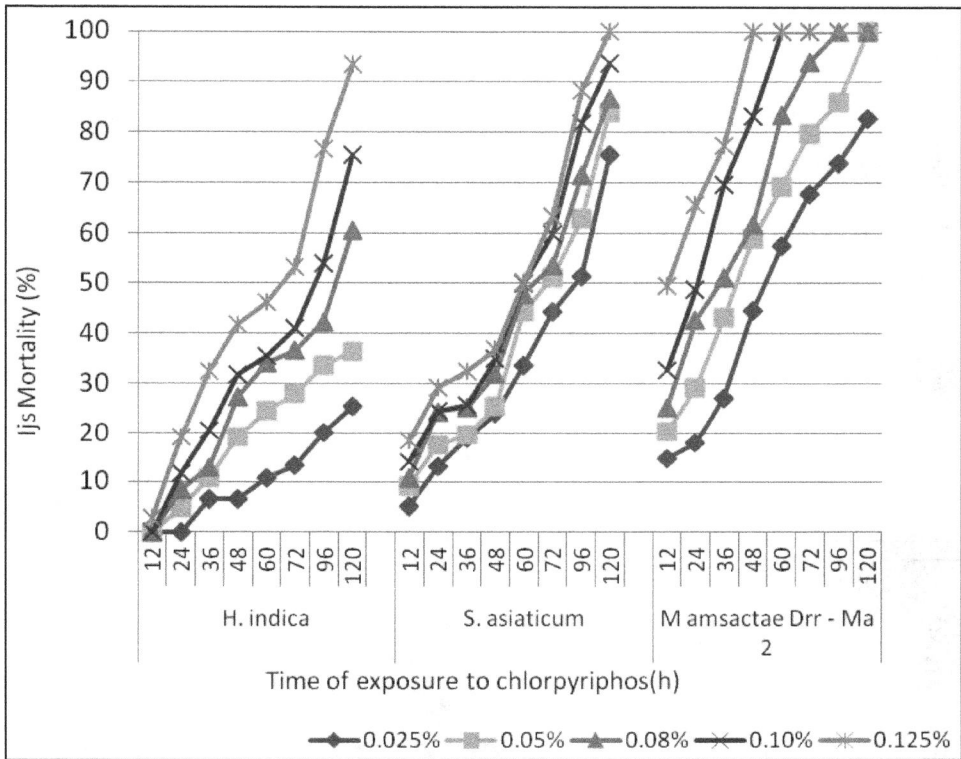

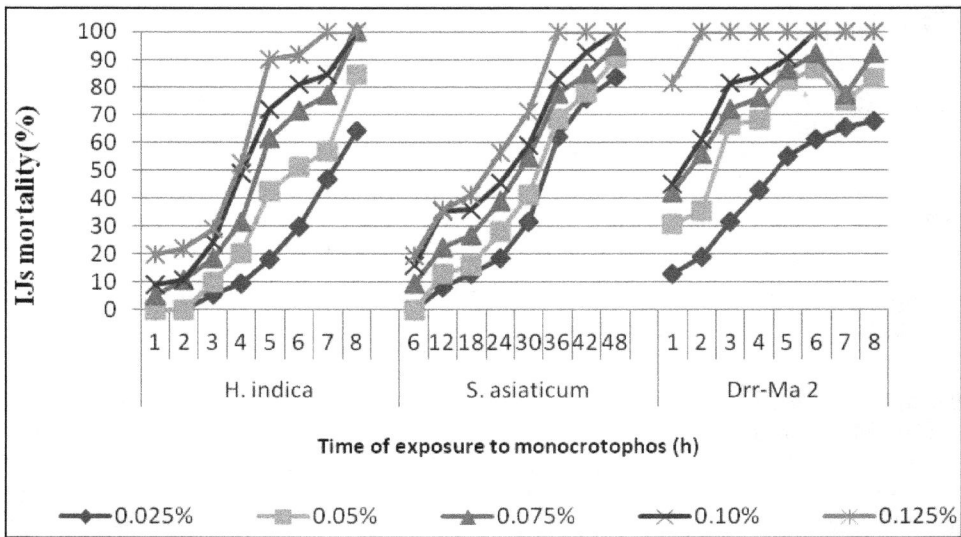

Contd...

Figures 14.1–*Contd...*

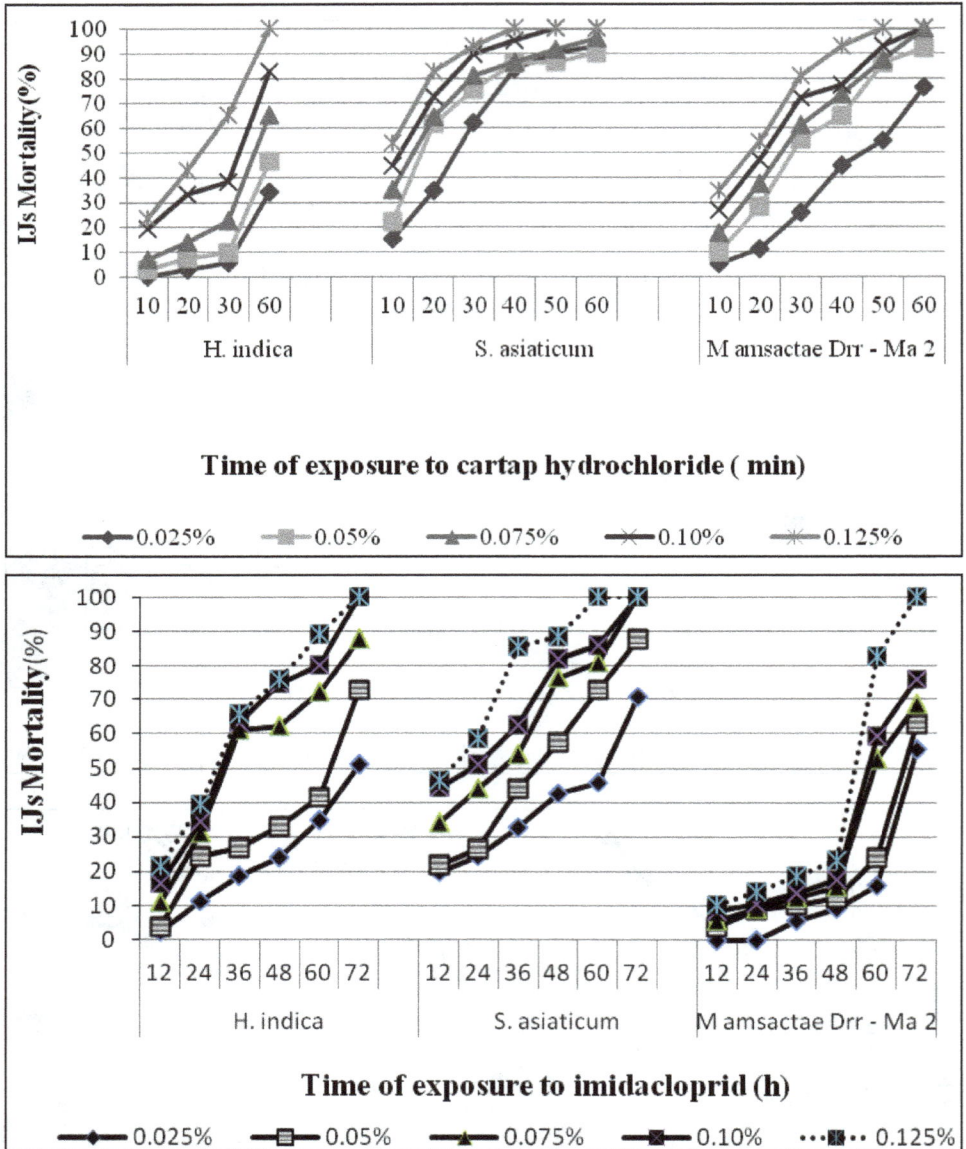

Time of exposure to cartap hydrochloride (min)

──◆── 0.025%   ──■── 0.05%   ──▲── 0.075%   ──✕── 0.10%   ──✱── 0.125%

Time of exposure to imidacloprid (h)

──◆── 0.025%   ──☐── 0.05%   ──▲── 0.075%   ──✕── 0.10%   ···■··· 0.125%

indica were infective though the time taken for larval mortality was as high as 106 h. Drr- Ma2 and *S. asiaticum* were sensitive to chlorpyrifos. *H. indica* IJs incubated with chlorpyrifos could infect and the time taken for infectivity was only 67.2 h. Of all the EPNs tested, *H. indica* showed maximum compatibility for use along with insecticides followed by *S. asiaticum* and *Drr -Ma2* (Table 14.1). The time taken for larval mortality of the IJs of *H. indica* and *S. asiaticum* varied from 24 to 48 h upto 0.075 per cent but with increase in concentration of insecticide, there was decrease in

Table 14.1: Effect of Adjuvants on EPN IJs

Mean per cent mortality of IJs

Triton X 100

Time of Exposure	M.amsactae Drr-Ma 2			S. asiaticum			H. indica		
	0.50 per cent	1 per cent	2 per cent	0.50 per cent	1 per cent	2 per cent	0.50 per cent	1 per cent	2 per cent
1h	2.7	4.0	5.3	3.7	5.3	6.7	2.6	3.0	4.2
2h	3.3	4.67	7.0	4.7	10.0	13.7	3.0	3.6	4.6
3h	6.7	9.33	12.7	12.7	15.3	24.0	3.4	4.6	6.4
4h	10.0	14.67	23.3	21.0	34.3	49.0	4.0	6.8	12.8
	Dose NS Time Fval 8.78 P≤ 0.0001 Dose x time NS			Dose NS Time Fval 3.17 P≤0.0333 Dose*Time NS			Dose Fval 177.97 P≤0.00000 Time Fval 143.63 P≤0.00000 D X t 22.2 P≤0.00000		

Tween 20

Time of Exposure	0.50 per cent	1 per cent	2 per cent	0.50 per cent	1 per cent	2 per cent	0.50 per cent	1 per cent	2 per cent
1h	2.0	3.3	5.3	2.7	2.7	11.3	7.2	12	17.4
2h	3.3	4.7	6.3	5.7	8.3	31.0	8.2	12.6	25.4
3h	6.7	9.9	12.7	6.3	7.3	36.0	8.6	13.8	29.2
4h	10.0	14.7	23.3	8.3	11.3	39.3	11.0	16.8	39.4
	Dose NS Time Fval 14.9829.35 P≤ 0.0046 Dose X time NS			Dose NS Time Fval 7.11 P≤0.0021 Dose*Time- NS			Dose Fval 742.52 P≤0.00000 Time Fval 91.64 P≤0.00000 D X t 17.62 P≤0.00000		

Contd...

Table 14.1–Contd...

Glycerol									
Time of Exposure	0.50 per cent	1 per cent	2 per cent	0.50 per cent	1 per cent	2 per cent	0.50 per cent	1 per cent	2 per cent
1h	0	0	2.33	5.3	11.3	14	4.6	7.6	6.6
2h	2.67	4.67	5.33	10.7	26	24.7	4.8	8.4	7
3h	7.33	9.33	12.67	13.0	34.7	46.0	5	8.8	7.6
4h	12.0	20.0	35.33	22.33	39.0	53.0	6.6	9.6	9.4
	Dose NS			Dose-NS			Dose - Fval 36.7 p ≤0.00		
	Time Fval 8.18 at			Time-NS			Time -Fval 8.14 p ≤0.0		
	$P \leq 0.0000$			Dose X Time-NS			Dose X time -NS		
	Dose X time NS								

Liquid Paraffin									
Time of Exposure	0.50 per cent	1 per cent	2 per cent	0.50 per cent	1 per cent	2 per cent	0.50 per cent	1 per cent	2 per cent
1h	4.3	8.7	11.3	7	15.3	18	4.6	7	10.4
2h	6	7.3	13	10	18.7	22	5.2	5.6	11.6
3h	11.3	14	20.3	13.3	21.3	28.3	9.8	12.4	16
4h	15.3	24.3	26.7	17.7	24.3	33.3	14	18.8	25
	Dose - NS			Dose - NS			Dose -F cal 36.7 p ≤0.00		
	Time - NS			Time- NS			Time - F cal 8.14 p ≤0.0		
	Dose X time NS			Dose X time NS			Dose X time NS		

infectivity. Based on the mortality of IJs, time taken for exposed IJs to cause mortality to *G. mellonella* and recovery of the EPNs, chlorpyrifos was found to be the safest insecticide and *H. indica* the most tolerant EPN. Zhang Li *et al.* (1994) reported that of all the insecticide groups tested, cartap hydrochloride was found to be highly toxic to *S. carpocapsae* IJs. Koppenhöfer and Fuzy (2008) reported the compatibility of imidacloprid with *Heterorhabditis bacteriophora*. Sivamuthuprakash *et al.* (2011) concluded that imidacloprid was safe but organophosphates (chlorpyrifos and monocrotophos) had adverse effect on *H. indica* IJs survival and infectivity. Pervez and Ali (2012) reported that when IJs were incubated with monocrotophos for 24 h there was reduction in infectivity, but there was variation in the extent of mortality of *Steinernema* sp. Nitin *et al.* (2013) reported compatibility of *S. carpocapsae* with imidacloprid. ŽigaLaznik and StanislavTrdan (2013) reported that compatibility with insecticides was species specific and strain specific.

Effect of UV light on EPN

The mortality of IJs of Drr EPN Ma 2 was 0.66 per cent after 20 minutes which suddenly increased to 39.7 per cent after 30 minutes of UV exposure. Cent per cent IJs mortality was observed at 150 minutes. In *S. asiaticum*, mortality started at 30 minutes with a minimum of 1.33 per cent and with increase in time of UV exposure IJs mortality reached cent per cent at 160 minutes (Figure 14.2). In case of *H. indica*, 87.5 per cent mortality of IJs was recorded at 16 minutes and 100 per cent mortality was observed after 19 minutes of exposure. The study showed that *H. indica* was very sensitive to UV radiation as compared to *Steinernema* sp. The surviving IJs of *S. asiaticum* after exposure to UV light for lesser time (5-10 minutes) showed faster pathogenicity causing 17 to 20 per cent mortality of host larvae (*G. mellonella*) within 24 h compared to 0 to 7 per cent mortality due to IJs exposed for longer duration (20-30 minutes). With increase in duration of exposure to UV radiation the infective

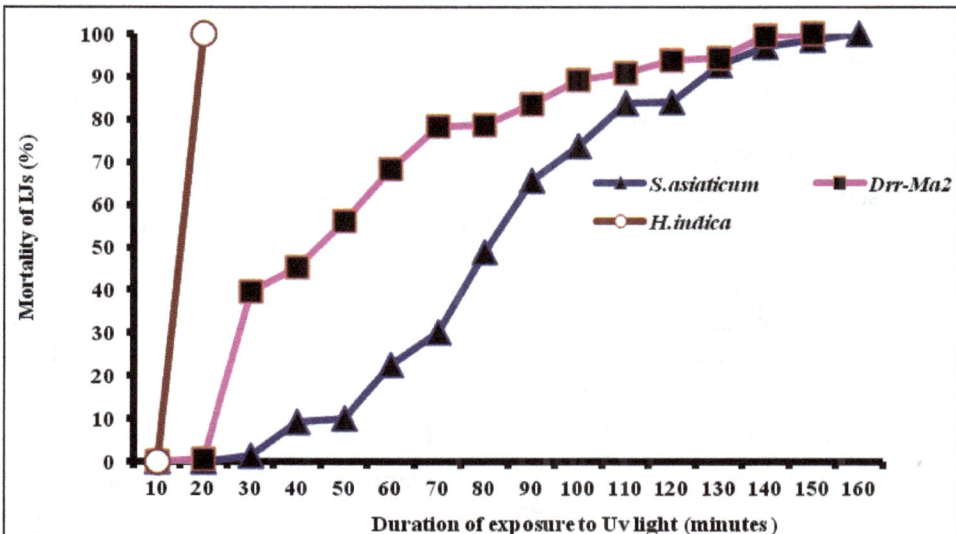

Figure 14.2: Effect of UV Light on Survival of EPN IJs.

capacity of the EPN decreased while there was increase in the time taken for host mortality. Ijs of *S. asiaticum* when exposed to UV light up to 70 min (96 h for larval mortality) could infect *G. mellonela* larvae. But there was loss of infectivity of *H. indica* IJs within 20 min. of exposure to UV light. The results revealed that *H. indica* was very sensitive to UV light as a short exposure of only 10 min to UV light reduced the infectivity and IJs took nearly 60 h to infect host larvae. IJs of DRR Ma 2 were more stable and there was no change in the time taken for larval mortality of *G. mellonella* when IJs were exposed upto 30 min. (48h) for UV light. Infectivity of IJs of Drr -Ma 2 was also affected on exposure to UV light for more than 40 minutes. Thereafter, there was increase in the time taken for larval mortality (88h for 50 min of exposure), which could be due to decrease in the infectivity of IJs. The recovery of IJs from the host cadavers was the highest in IJs exposed for 5 minutes (51763/larva) and showed a declining trend with increase in duration of UV exposure. The recovery was the lowest (13446/larva) in IJs exposed for longest duration of 30 minutes. Hussaini *et al.* (2003) observed that as the exposure to UV light increased there was an increase in the mortality of IJs.

Effect of Adjuvants on Survival and Infectivity of EPN

Experiments showed that mortality of IJs increased with the increase in time of incubation of IJs with adjuvants, and concentration of the IJs, mortality of the IJs increased. Time of exposure had significant effect on mortality of IJs in all the combinations of adjuvant and EPN sp except in case of liquid paraffin on *Steinernema* sp. and glycerol on *S. asiaticum* (Table 14.1). Dose, time and their interaction effect were significant in case of Triton X 100 on *H. indica*. The lower dose of 0.5 per cent was safer to all the EPNs tested. Consequently, there was a decrease in the infectivity of IJs towards *G. mellonella*. Among the EPNs, *S. asiaticum* was more sensitive. The surfactants, Labklin R and labolene R when added to the EPN suspension, were observed to be toxic to IJs. Hussain *et al.* (2005) reported that Triton X 100 and Tween 80 protected the IJs from desiccation at higher concentrations of upto 2 per cent when incubated for 1 and 2 h only without anyadverse effect. Glycerol, triton X 100 and Tween 20 had lower IJ mortality at 0.5 and 1 per cent concentrations and offered maximum protection and could be utilized as effective antidesiccants in foliar sprays of nematodes.

Based on pathogenicity of exposed EPN IJs to *G. mellonella* larvae, Triton X 100 had least effect on EPNs, showing faster host larval mortality compared to glycerol and Tween 20 which were on par and better than liquid paraffin. Among the EPNs, *H. indica* was least affected as it resulted in faster host larval mortality followed by Drr-Ma 2 and *S. asiaticum*. The data on recovery of IJs revealed that liquid paraffin solution showed highest recovery of *S. asiaticum* and Drr-Ma 2, while glycerol solution showed highest recovery of *H. indica*.

Efficacy of EPN in Greenhouse

Spray application of *Drr-Ma 2* to potted plants resulted in significantly lower dead heart (DH) damage (7.2 per cent) compared to 16.0 per cent DH in control pots. *Drr–Ma*1 treated pot also showed less stem borer damage (24.4 per cent DH) than control (31.8 per cent DH). A spray of either *S. asiaticum* and *S. carpocapsae* in the nursery resulted in 23.7 per cent and 28.0 per cent dead hearts, respectively showing 55-60 per cent reduction in damage in the nursery as compared to control (64.1 per cent DH) (Table 14.2).

Table 14.2: Efficacy of EPN Species on Yellow Stem Borer Damage in Greenhouse at Vegetative Phase

Treatment	Dead Hearts (Per cent)		
	Drr- Ma1	Drr -Ma2	S.asiaticum
Treated	24.4	7.2	28.0
Control	31.8	16.0	64.1
t value p=0.05	3.20** at 6 d.f	2.78*at - 9 d.f	3.73**at 11df

Evaluation of the Efficacy of EPN in Field

EPN treated plots showed upto 8.8 per cent white ears as compared to 12.5 per cent in the control and 5.2 per cent in the cartap hydrochloride treatment in *Kharif* 2007. In *Rabi* 2007 also, the damage varied from 3.3-4.3 per cent in EPN treated plots as compared to 1.2 per cent in cartap treatment and 5.3 per cent in water treatment (Table 14.3). Our studies also indicated the recovery of the EPN from the egg mass of YSB even 10 days after field application. Spraying of EPN in limited area across years with various species against yellow stem borer resulted in 23.9 to 37.7 per cent reduction in white ear damage as compared to control. In cartap hydrochloride treatment 54.9 per cent reduction in white ear damage was observed (Table 14.4). The grain yield was significantly higher in the EPN treatments (0.49-0.5kg/m²)as compared to control (0.40kg/m²) though insecticide treated plots recorded higher yield (F val 4.17 at 43 df; P= 0.0066). Studies in West Africa revealed that field application of *H. indica* on rice could reduce the damage by *Maliarpha separatella*, the African white stem borer (Kega *et al.*, 2013).

Table 14.3: Effect of EPN Spray on Yellow Stem Borer Damage in Field

EPN Treatments	White Ear Damage (Per cent)	
	Kharif 2007	Rabi 2007
H.indica	8.1ab	NT
Drr- Ma1	NT	3.3ab
Drr -Ma2	8.8ab	4.3b
cartap	5.8a	1.2a
Control	12.8b	4.5b
water	10.9b	5.3b
F value	2.76 * at 28d.f	3.35 at 36 d.f

Values in a column followed by different letters are significant at P≤0.05, NT-Not tested.

Table 14.4: Field Efficacy of EPN Species on Yellow Stem Borer Damage

Season of Study	EPN sp.	Variety	Per cent Reduction in WE Damage by YSB Over Control
Rabi 2007	M.amsactae Drr Ma1	Rasi	37.7
Rabi 2007	M.amsactae Drr -Ma 1	Jarava	34.3
Kharif 2007	H.indica	Rasi	36.3
Kharif 2007	M.amsactae Drr -Ma 2	Rasi	31.4
Rabi 2010	M.amsactae Drr -Ma2	TN1	24.4
Rabi 2010	M.amsactae Drr Ma1	TN1	23.9
Rabi 2010	S.asiaticum	TN1	41.5
Kharif 2007	Cartap hydrochloride	Rasi	54.9

Conclusions and Future Prospects

Our studies indicate that the indigenous EPN isolates have the ability to locate the host insects in rice plant and infect them. The species varied in their tolerance to UV rays and compatibility with adjuvants and insecticides. EPNs could significantly reduce the damage by yellow stem borer at all the stages of the crop as they could infect egg, larvae and pupae. The studies indicate that EPN are effective bioagents and hold a great promise if their application is well timed to target the vulnerable stages which are not amenable for control by insecticides. They can be incorporated in IPM to offset the biology and development of the pest so as to keep the populations below the threshold level. But care must be taken to apply them in the evenings so as to minimize the exposure to sunlight and not use in combination with insecticides like cartap hydrochloride or imidacloprid.

Acknowledgements

The authors thank Project Director, IIRR for his unstinted support and part of the work was carried out under BT/PR4068/AGR/05/225/2003 funded by DBT, Government of India.

References

Campbell J F, Orza G, Yoder F, Lewis E and Gaugler R 1998. Spatial and temporal distribution of endemic and released entomopathogenic nematode populations in turfgrass. *Entomologia Experimentalis et Applicata* **86:** 1-11.

Devi Gitanjali 2011. Influence of agrochemicals and botanicals on the mortality of entomopathogenic nematodes (Meghalaya isolates). *Indian Journal of Nematology* **41:** 127-130.

Dhaliwal G S, Jindal Vikas and Dhawan A K 2010. Insect pest problems and crop losses: Changing trends. *Indian Journal of Ecology* **37:** 1-7.

Duncan L W, Graham J H, Dunn D C, Zellers J, McCoy C W and Nguyen K 2003. Incidence of endemic entomopathogenic nematodes following application of *Diaprepes abbreviatus*. *Journal of Nematology* **35:** 178-186.

Grewal P S and Georgis R 1999. Entomopathogenic nematodes.*In* "Methods in biotechnology", Vol 5, "Biopesticides: use and delivery" (F. R. Hall and J. J. Menn, Eds.), pp 271-299. Humana Press Inc., Totowa.

Gururaj Katti, Prasad J S, Padmakumari A P and Sankar M 2006. Effect of storage period on survival and infectivity of indigenous entomopathogenic nematodes of insect pests of rice. *Nematologia Mediterranea* **34:** 37-41.

Hussaini S S, Singh S P and Shakeela V 2001. Compatibility of entomopathogenic nematodes (Steinernematidae, Heterorhabditidae: Rhabditida) with selected pesticides and their influence on somebiological traits. *Entomon* **26:** 37-44.

Hussaini S S, Hussain M A and Satya K J 2003. Survival and pathogenicity of indigenous entomopathogenic nematodes in different UV protectants. *Indian Journal of Plant Protection* **31:** 12-18.

Hussaini S S, Nagesh M, Rajeshwari R and Dar M H 2005. Effect of antidesiccants on survival and pathogenicity of some indigenous isolates of entomopathogenic nematodes against *Plutella xylostella* (L.). *Annals of Plant Protection Sciences* **13:** 179 -186.

Katti G, Padmakumari A P and Prasad J S 2003. An entomopathogenic nematode infecting rice yellow stem borer, *S. incertulas. Indian Journal of Plant Protection* **31:** 80-84.

Kega V M, Kasina, Molubayo and Nderitu 2013. Management of *Maliarpha separetella* Rag using effective entomopathogenic nematodes and resistant rice cultivars. *Journal of Entomology* **10:** 103-109.

Koppenhöfer A M and Fuzy E M 2008. Early timing and new combinations to increase the efficacy of neonicotinoid-entomopathogenic nematode (Rhabditida: Heterorhabditidae) combinations against white grubs (Coleoptera: Scarabaeidae). *Pest Management Science* **64:** 725-35. doi: 10.1002/ps.1550.

NitinKulkarni, Sanjay Paunikar and Vinod Kumar Mishra 2013. Tolerance of entomopathogenic nematode *Steinernema carpocapsae* to some modern insecticides and biopesticides. *Annals of Entomology* **31:** 129-134.

Padmakumari A P, Katti G, Sankar M and Prasad J S 2005. Efficacy of entomopathogenic nematodes on rice yellow stem borer, Scirpophaga incertulas. Contributions to National Seminar on Biotechnological Management of Nematodes and Scope of Entomopathogenic Nematodes, held at Sun Agro Biotech Research Centre, Chennai during November 21-22, 2005. Sun Ray No.1. 185 pp. Eds. Sithanantham, S., Vasanthraj David, B. and Selvaraj, P. pp. 172-174.

Padmakumari A P, Prasad J S, Gururaj Katti and Sankar M 2007. *Rhabditis* sp. (*Oscheius* sp.).a biocontrol agent against rice yellow stem borer, *Scirpophaga incertulas. Indian Journal of Plant Protection* **35:** 255-258.

Padmakumari A P, Gururaj Katti, Sankar M and Prasad J S 2008. Entomopathogenic nematodes in rice pest management. Technical Bulletin No.32. Directorate of Rice Research, Rajendranagar, Hyderabad-500030, A.P., India.21 pp.

Parkman J P, Frank J H, Nguyen K B and Smart G C Jr 1993. Dispersal of *Steinernema scapterisci* (Rhabditida: Steinernematidae) after inoculative applications for mole cricket (Orthoptera: Gryllotalpidae) control in pastures. *Biological Control* **3:** 226-232.

Pervez Rashid and Ali S S 2012. Compatibility of entomopathogenic nematodes (Nematoda: Rhabditida) with pesticides and their infectivity against lepidopteran insect pest. *Trends in Biosciences* **5:** 71-73.

Prasad J S, Katti G, Padmakumari A P and Pasalu I C 2003. Exploitation of indigenous entomopathogenic nematodes against insect pests of rice. Pp. 121-126. *In*: Current status of research on entomopathogenic nematodes in India (Hussaini S.S., Rabindra R.J. and Nagesh M., eds.). Project Directorate of Biological Control, Bangalore, India.

Sankar M, Prasad J S, Padmakumari A P and Katti G 2009. Efficacy of entomopathogenic nematode, *Heterorhabditis indica* against the rice yellow stem borer, *Scirpophaga incertulas* (Wlk.) (Lepidoptera: Pyralidae). *Biopesticides International* **5:** 68-74.

Shapiro-Ilan D I and Gaugler R 2002. Production technology for entomopathogenic nematodes and their bacterial symbionts. *Journal of Industrial Microbiology and Biotechnology* **28:** 137-146.

Sivamuthuprakash M, Varadarasan S and Jayanthi Abraham K 2011. Compatibility studies of *Heterorhabditis indica* (ICRI-18) with commonly used pesticides and fungicides at cardamom plantations. *Ecology, Environment and Conservation Paper* **17:** 563-566.

Zhang li, Shono Toshio,Yamanaka Satoshi and Tanabe Hiroshi 1994. Effects of insecticides on the entomopathogenic nematode, *Steinernema carpocapsae* Weiser. *Applied Entomology and Zoology* **29:** 539-547.

Žiga Laznik and Stanislav Trdan 2013. The influence of insecticides on the viability of entomopathogenic nematodes (Rhabditida: Steinernematidae and Heterorhabditidae) under laboratory conditions. *Pest Management Science* **70:** 784-789.

Chapter 15

Indian Scenario of Invasive Mealybug, *Paracoccus marginatus*

**M Mani[1]*, A N Shylesha[2], A Krishnamoorthy[1],
M Kalyanasundaram[3], R V Nakat[4], K R Lyla[5]
and S J Rahman[6]**

[1]*ICAR-Indian Institute of Horticultural Research, Bengaluru – 560 089*
[2]*ICAR- National Bureau of Agricultural Insect Resources,
Bengaluru – 560 024, Karnataka*
[3]*Tamil Nadu Agricultural University, Coimbatore – 641 003*
[4]*Maharashtra Krishi Vidyapeeth, Rahuri – 413 722, Maharashtra*
[5]*Kerala Agricultural University, Thrissur – 680 656, Kerala*
[6]*Professor Jayashankar Telangana State Agricultural University,
Hyderabad – 500 030, Telangana State*
E-mail: mmani1949@yahoo.co.in

ABSTRACT

The papaya mealybug (PMB), Paracoccus marginatus Williams and Granara de Willink (Hemiptera: Pseudococcidae) is native of Mexico and/or Central and North America. Since its description in 1992, it has invaded many countries including India. It was first found at Coimbatore in July 2008 in Tamil Nadu and subsequently spread to Karnataka, Andhra Pradesh, Maharashtra, Kerala, Tripura, Assam, Orissa and of late in Lakshadweep islands, Gujarat, Rajasthan and New Delhi. It is highly polyphagous attacking more than 85 plant species causing severe loss. Papaya, mulberry, tapioca and cotton are the economically important crops affected seriously with the mealybugs.Insecticides failed to give adequate control of P. marginatus. On the other hand, natural enemies particularly Acerophagus papayae Noyes and Schauff was highly useful to suppress the papaya mealybug in many countries. The parasitoid was obtained from USDA-APHIS Puerto Rico by NBAIR, Bengaluru, during July-October, 2010. After ascertaining the safety in quarantine, the parasitoid was distributed to different states in India. The parasitoid was multiplied on large scale and released in the farmer's fields of papaya in several districts. There was substantial reduction of PMB density to very low level within three months of its introduction. In depth details on the state-wise occurrence and spread in the country, and management of this invasive pest using parasitoids are discussed in this paper.

Introduction

Mealybugs throughout the world cause a variety of economic problems. *Paracoccus marginatus* Williams and Granara de Willink (*Hemiptera: Pseudococcidae*), popularly known as papaya mealybug (PMB) is native of Mexico and/or Central and North America andhas invaded several countries and damaged many economically important crop plants. Papaya, mulberry, tapioca and cotton are the economically important crops affected seriously with the mealybugs. It is 'hard to kill pest' with conventional insecticides because of protected habitat and waxy coating over the body. Classical biological control with the exotic parasitoid, *Acerophagus papayae* Noyes and Schauff has given excellent control of *P. marginatus* in several countries including India (Meyerdirk *et al.*, 2004; Muniappan *et al.*, 2006; Galanihe *et al.*, 2010; Mani *et al.*, 2012).

P. marginatus is native to Mexico and/or Central and North America (Miller and Miller 1999). Since its first description in 1992 from new tropical region, *P. marginatus* has spread to several countries (Mani *et al.*, 2012). It was noticed in South and South East Asian region during 2008-2010. It was recorded in Indonesia, Philippines and Sri Lanka in 2008; In India, it was found at Coimbatore in July 2008 in Tamil Nadu (Muniappan *et al.*, 2009; Regupathy and Ayyasamy, 2009; Suresh *et al.*, 2010). Since July 2008, it has spread from Coimbatore in Tamil Nadu, to other states such as Karnataka, Andhra Pradesh, Maharashtra, Kerala, Tripura, Assam, Orissa, Rajasthan, Gujarat and New Delhi in India (Krishnakumar and Rajan, 2009; Lyla and Philip, 2010; Mahalingam *et al.*, 2010; Rabindra, 2010; Suresh *et al.*, 2010; Chandele *et al.*, 2011; Jacob Mathew, 2011; Krishanamurthy and Mani 2011; Mani Chellappan, 2011; Sajeev, 2011; Shylesha *et al.*, 2011b).

Damage

P. marginatus attacks over 85 species of plants including field crops, fruit trees ornamentals, weed and scrub vegetation in India. Papaya mealybug infestations are typically observed as clusters of cotton-like masses on the above-ground portion of plants. *P. marginatus* damages various parts of the host plant including the leaves, stems, flowers and fruits. *P. marginatus* showed very similar symptoms to pink hibiscus mealybug, *Maconellicoccus hirsutus* (Green). The insect sucks the sap by inserting its stylets into the epidermis of the leaf, fruit and stem. While feeding, it injects a toxic substance into the leaves resulting in curling, crinkling, rosetting, twisting and general leaf distortion. Heavy mealybug infestations render fruit inedible. Due to the build-up of thick white waxy coating and sooty mould development on the honeydew excreted by mealybug, infested fruits get reduced market value. Fruits may fail to develop normally and may be unusually small. Such fruits eventually shrivel and drop. Some economically important crops damaged by *P. marginatus* are papaya, mulberry, cotton, cassava, citrus, avocado, guava, jackfruit, sweet potato, peas and beans, okra, eggplant, and ornamentals such as hibiscus, *Jatropha, Allamanda, Acalypha* (Shylesha *et al.*, 2011). Some common weeds like *Parthenium* sp., *Althea* sp., *Sida* sp., *Amaranthus* sp. etc are highly preferred.

Biological Control

Though several methods were available, excellent control of *P. marginatus* was achieved with parasitoids in several countries (Mani *et al.*, 2012). The import and quarantine of the papaya mealybug parasitoids is the first case after the state of the art quarantine building has been set up by the ICAR at National Bureau of Agricultural Insect Resources (NBAIR). The timely funding of the project on papaya mealybug in mulberry by *Central Sericultural Research* and *Training Institute* (CSRTI) had boosted the biological control programme. The NBAIR has successfully imported three species of parasitoids, *A. papayae, Anagyrus loecki* Noyes and *Pseudleptomastix mexicana* on 15th July 2010 with the help of United States Department of Agriculture-Animal and Plant Health Inspection Services (USDA-APHIS) from their facility at Puerto Rico and completed all the mandatory safety and specificity tests in the quarantine facility. The NBAIR has mass multiplied these parasitoids successfully in the quarantine laboratory on *P. marginatus* colonies grown on potato sprouts. Filed release of the parasitoids was initiated on the 7th of October 2010 in Tamil Nadu followed by other states.

Status of *Paracoccus marginatus* in differerent States

Tamil Nadu

P. marginatus was first observed on 10th July 2008 in papaya plantations located in and around Coimbatore and subsequently spread to neighboring districts of Tamil Nadu. In addition to papaya, this mealybug affected silk industry by attacking the mulberry plant, Sago industry by attacking tapioca plant, bio fuel industry by attacking jatropha and timber industry by attacking teak. There was 75 per cent damage on papaya, 65 per cent on tapioca, 60 per cent on mulberry, 30 per cent on cotton, 25 per cent on pulses chiefly pigeon pea, 90 per cent on jatropha, 65 per cent on vegetables including okra and 90 per cent on silk cotton caused by *P. marginatus* in 2009-10.

Papaya growers who are cultivating papaya for papain suffered huge losses due to the mealybug attack. Looking into the severity, some farmers had opted for uprooting papaya trees and some commercial growers have abandoned papaya cultivation. Intially, the papaya crop of worth 80 crores was lost due to mealybug attack. Chemicals like monocrotophos, methyl demeton, dimethoate, acephate, methomyl, fenthion, imidacloprid, thiomethoxam, dichlorovos, quinalphos, profenophos, fenitrothion, carbaryl, chlorpyriphos, diazinon, malathion, buprofezin were used against papaya mealybug desperately.They gave short-term control but could not solve the mealybug problem. In October 2010, the parasitoid, *A. papayae* was obtained from NBAIR, Bengaluru, and multiplied on large scale and released in the farmer's field of papaya in several districts in Tamil Nadu. There was substantial reduction of PMB density to very low level within three months after its introduction (Jonathan *et al.*, 2011; Kalyanasundaram *et al.*, 2011). Pooled analysis on the results obtained from the field experiments conducted in different locations in Tamil Nadu revealed that there was 8 per cent, 32 per cent, 65 per cent, 80 per cent and 96 per cent reduction in the mealybug population on 15,30,45,60 and 90 days after the release of the parasiotoid. There was a monetary benfit of Rs. 360 lakhs to papain industry alone in 2011-12 and 540 lakhs in 2012-13 by way of controlling the mealybugs through the release of *A. papayae* (Regupathy and Ayyasamy, 2011).

P. marginatus was first noticed on mulberry in Avinsahi taluk of Coimbatore district during January 2009. The infestation had spread very fast in Coimbatore and Erode districts by March 2009, infesting nearly 1500 to 2000 acres of mulberry, and the farmers could not take silkworm rearings as the brushing capacity of gardens was reduced by 80 to 90 per cent. Later it had spread to Namkkal, Salem, Trichy and Dharmapuri districts. There was 60 per cent damage by the mealybug on mulberry in T.N. A total area of 10,000 acres of mulberry gardens was found infested with *P. marginatus*. It was estimated that mulberry crop worth of 135 crores was lost due the mealybug infestation in T.N. The local predators, *viz.*, *Spalgis epeus* Westwood, *Cryptolaemus montrouzieri* Mulsant and *Scymnus coccivora* Ayyar were found feeding on the mealybugs under natural condition but unable to check the mealybug damage. According to Qadri *et al*. (2011), more than 33,000 adults of *A. papayae* were released (from Nov'10 to March'11) in the mealybug infested mulberry gardens of 350 farmers in the districts of Erode, Tiruppur and Salem. *A. papayae* was found most effective and aggressive, and same was released in papaya mealybug infested mulberry gardens @ 100 adults per acre. Subsequently, the chemical application for the papaya mealybug as recommended in mulberry was totally discontinued. After the release of the parasitoids, the mealybug infestation was reduced from 90 per cent to less than 5 per cent thereby achieving a suppression of 85-95 per cent within six months of release. Similiar control was achieved with the parasitoid in Trichy and Coimbatore districts in Tamil Nadu by April 2011, the incidence of the pest in mulberry drastically reduced to < 5 per cent. Simultaneously brushing capacity also came to normal by mid 2011-12.To determine the impact of papaya mealybug attack on brushing capacity of mulberry garden, an index value of 100 was considered for the year 2008-09 (ie. prior to pest attack). After the attack of the pest, index value was 85.37 for the year 2009-10 and 65.89 for 2010-11. However, the index value sharply raised to 98 during 2011-12 within six months of release of exotic parasitoid, *A. papayae*. The outcome of classical biological control programme has given rich dividends and the papaya mealybug infestation was reduced to bare minimal levels enabling the sericulturists to improve the brushings to >90 per cent. It reversed the tendency of farmers to uproot the mulberry plantation and revived the sericultural activities in severely affected areas and ushered a new hope among the sericulturists to fight against this dreadly pest successfully.

The mealy bug was reported on cassava in Tamil Nadu during 2009 causing severe damage. Heavy infestation caused leaf shedding and yield loss. The mealybug infestation varied from 50-90 per cent in cassava resulting in a monetary loss of rupees 220 crores in cassava alone in Tamil Nadu. Severe infestation of *P.marginatus* was observed on cassava in Namakkal, Salem and Dharmapuri districts of Tamil Nadu state (Sakthivel and Qadri, 2010) besides Coimbatore, Karur, Erode, Thirupur and Trichy districts. Heavy population load @ 38.70, 43.85 and 41.21 numbers/5cm^2 was recorded in Salem, Dharmapuri and Namakkal districts, respectively. Releases of *A. papayae* were made @ 200 individuals per location in Tamil Nadu in 2010. The mealybug population had declined uniformly corresponding to gradual increase in percent parasitism. The average population of papaya mealybug from the tapioca gardens was eliminated up to 93.15 per cent at 6th month. Parasitism by *A. papayae* on the mealybug accounted for 80 - 95 per cent per cent (Sakthivel, 2013). Similar control of the mealybug was achieved with the release of *A. papayae* in other districts *viz.*, Trichy and Erode (Divya 2012) and Karur (Vijay, 2010) in Tamil Nadu, India.

Cotton

A. papayae was found highly effective against on *P. marginatus* on cotton in and around Coimbatore (Dharajothi *et al.*, 2011).

Karnataka

As many as 85 plant species including papaya, mulberry, teak, plumeria. Marigold, Jatropha, areca, Ixora, Nerium etc. were found infested with *P. marginatus* in Karnataka in 2010. *A. papayae* was multiplied in large numbers, and released initially on Jatropha in Bengaluru during October 2010 for establishment. Susequently, *A. papayae* was released in GKVK, Jakkur, Mysore, Chamaraj Nagar, Mandya, Hesaraghatta, Nelamangala, Gulbarga, Shiddalaghatta, Bagalur, Kollegal in 2010-11. Recovery of *A. papayae* was made in few numbers on 20th day after release, and later in very large numbers on 40 and 60th day of release. After a span of three months, there was a reduction of 80-90 per cent in the mealybug population, and new shoots developing were not found to harbor any mealybug (Shylesha *et al.*, 2011b; Krishanamurthy *et al.*, 2011; Qadri *et al.*, 2011).

Mulberry as well as other crop plants were found attacked by *P. marginatus* in Chamarajanagar district during mid 2010 in villages such as Honnur, Honganur, Gangavadi, Byadamudlu (Yelandur Taluk), Dollipura, Hanahalli, Mookahalli, Yediyur (Chamarajanagar Taluk), Hanumanapura and other villages (Nanjungud Taluk) with an initial infestation ranging from 60 to 90 per cent. Subsequently, spread to neighboring districts of Mysore, Mandya and to some extent Ramanagara districts. A total of 23,750 adults of *A. papayae* were released in the mealybug infested gardens of 91 farmers covering about 181 acres of mulberry in RSRS, Chamarajanagar. In addition, CSRTI, Mysore had released a total of 27,000 parasitoids covering 70 acres in Annur area (Maddur Taluk). The parasitoids were also released in infested mulberry gardens located in the districts of Mysore, Mandya, Ramanagar, Raichur and Chamarajanagar. After the release of parasitoids, the mealybug infestation was reduced by 80-90 per cent within 5 to 6 months of release saving the mulberry crop and thereby increasing the cocoon production. This has resulted in savings to the tune of few crores of rupees in Karnataka.

Papaya plantations were found affected with *P. marginatus* very seriously in April, 2010 in Bengaluru, Chamaraja Nagar, Mysore, Mandya, Gulbarga, Belgaum and Tumkur. By May-June 2010, mealybug infestation had spread across Bengaluru city and moved to farmers' fileds. Initially, about 25 per cent plants and later at the end of June almost over 60 per cent of the plants were found damaged by the invasive mealybug. Despite repeated chemical sprays with imidacloprid, profenofos, acetomiprid, dimethoate and buprofezin, the mealybug could not be controlled. The extent of fruit loss initially was about 50 per cent but subsequently it was between 80-100 per cent. *A. papayae* was released in papaya gardens located in and around Bengaluru, Hessaraghatta, Sidalaghatta, Gulbarga and several other places in Karnataka. Results in the experimental fields showed that the local predators like *S. epeus, C. montrouzieri* and *Scymus* sp. were found feeding on the mealybugs in negligible numbers. Sixty per cent of plants were found damaged in the grade of 4.0 in Sepetmber,2011(prior to the release of the papasitoids), and papaya plants were completely cleared of mealybugs (grade 0) by December. As many as 1100 parasiotoids were recovered from the samples collected in the third month of parasitoid release.

Other Host Plants

During October 2010, *A. papayae* was released on jatropha and plumeria in Bengaluru. After a span of 3 months, there was a reduction of 80-90 per cent in the pest population on these plants (Shylesha *et al.*, 2011b).

Maharashtra

P. marginatus was reported first in Pune in July, 2010, and later the mealybug had assumed the status of major pest of papaya in Dhule, Nandurbar, Jalgaon, Ahmednagar, Pune, Solapur, Satara, Sangli, Aurangabad, Jalna, Beed, Osmanabad, Nanded, Parbhani, Hingoli, Buldhana, Akola and Wasim districts of the state. Besides papaya, the mealybug was recorded on teak, guava, mulberry, sapota, *Amaranthus, Acalypha, Hibiscus, Plumeria* and several several weed plants. The local predators *viz., Spalgis epeus, Coccinella septempunctata* Linn, *Scymnus coccivora* Ayyar, *Mallada* sp., *Brumoides* sp., unidentified syrphids, anthocorids and spiders were found feeding on the mealybugs. The parasitoid *A. papayae* was found parasitising for the first time in August, 2010. About 8,000 parasitoids were produced during August to October, 2010 for making field release. Meanwhile, *A. papayae*, imported by the

Table 15.1: Effect of release of *A. papayae* on Incidence of PMB

Location (Dist.)	Pre-release PMB Incidence (per cent)	Per cent Decline in PMB Population*		
		1 MAR	2 MAR	3 MAR
P-I (Pune)	90.0 (5)	20.0	48.0	85.0
P-II (Pune)	95.0 (5)	24.0	56.0	92.0
P-III (Jalgaon)	85.0 (5)	8.0	20.0	38.0
P-IV (Dhule)	65.0 (4)	4.0	21.0	40.0
P-V (Thane)	65.0 (4)	12.0	35.0	70.0

*Decline in pest population computed on the basis of pest intensity rating.

Figures in bracket are pest intensity rating (1 = very low; 2 = low; 3 = Medium; 4 = High; 5 = Very high population).

MAR: Month after release; P: Papaya orchard.

NBAIR, Bengaluru was also brought to Pune in November, 2010 for mass culturing on potato sprouts. Field release of the parasitoid, *A. papayae* was undertaken during 2010-11, 2011-12 and continued in 2012-13 (Nakat *et al.*, 2011).

A release rate of 1000-1500/ac was adopted in Maharastra. There was 85-92 per cent decline in the mealybug population within three months of parasitoid release (Pokharkar *et al.*, 2011; Mundale and Nakat, 2011; Chandale *et al.*, 2011; Nakat *et al.*, 2011).

Table 15.2: Survey of Papaya Mealybug in different Agro-Climatic Zones of Western Maharashtra

Area/District Surveyed	2010-11	2011-12	2012-13
	PMB incidence (PIR) (Per cent)	PMB incidence (PIR) (Per cent)	PMB incidence (PIR) (Per cent)
Pune	95 (5.0)	21.5 (2.0)	35 (2.9)
Jalgaon	90 (4.5)	24.4 (2.2)	28 (2.7)
Dhule	80 (4.0)	28.5 (2.7)	30 (2.8)
Nandurbar	95 (5.0)	27.4 (2.6)	38 (3.0)
Thane	65 (4.0)		36 (3.0)
Kolhapur	60 (3.0)	6.0 (1.0)	–
Ahmednagar	–	–	14.0 (1.2)
Average	80.8 (4.1)	21.6 (2.1)	30.2 (2.6)

PIR- Pest intensity rating Gr.1-5.

Andhra Pradesh

P. marginatus was first reported in Kadapa and Chittor districts during 2010 followed by Anantapur in June 2012. The mealybugs were found damaging papaya, custard apple, cotton, tapioca, okra, redgram, brinjal, marigold, *Parthenium* sp., *Acalypha indica* etc. Releases of *A. papayae* were made on 14-10-2010 in papaya gardens in Kadapa and 17-8-2012 in Ananthapur district. Prior to parasite release, the crop damage was in the grade 5. There was 15, 45 and 90 per cent decline in the mealybug population damage on 30, 60, 90 days after release by the end of November, complete control of the mealybug was achieved in Ananatapur. In Kadapa and Chitoor districts, the papaya mealybug infestation was reduced by 80-92 per cent. A benefit of Rs 5 lakhs was obtained from five acre plot cultivated/ infested papaya field by releasing the parasitoid. In Medak district, the parasitoid was released in 18 acres planted with papaya. The crop worth of Rs 18 lakhs was saved with the release of the parasitoid. Currently average benefit per acre was: 1.5 Lakhs and the expected savings was to the tune of 135 crores in Andhra Pradesh.

Kerala

P. marginatus was first observed in Kerala during 2009. Papaya, mulberry, tapioca, brinjal, tomato, cowpea, ashgourd, guava,dolichos jack, plumeria, hibiscus, ocimum, teak, rubber and parthenium were found infested by *P. marginatus*. Papaya is cultivated only in homesteads. During October 2010 in different districts of Kerala, above 60 per cent of the papaya plants were infested with the mealybug, and the intensity of damage ranged from medium to very high (Grade 3 to 5). Prior to release of parasitoid in Kerala, application of one of the insecticides, *viz.*, profenophos, dichlorvos, acephate, imidacloprid, thiomethoxam, acetamipride along with neem oil emulsion was recommended. Both imidacloprid and profenophos were widely used against the papaya mealybug with the residues lasted for more than three weeks.The re-infestation had occurred within a week. Releases of *A. papayae* were made first at Malappuram on 09-12-2010. A release rate of 25-50 parasitoids/ homestead was adopted to control the papaya mealybug.

Table 15.3: Status of Papaya Mealybug in Kerala in 2012

District	No. of Villages	Mealybug Infestation (per cent)	Infestation Grade
Thrissur	10	2.2	Low- Medium
Ernakulam	4	3.7	Low – Medium
Palakkad	4	1.6	Very low
Malappuram	6	3.1	Very low
Wayanad	6	1.5	Very low
Kozhikkode	4	1.6	Low

Kerala has about 60 lakh home steads and papaya is cultivated in these gardens.S avings to the tune of Rs. 72 crores were realised by controlling the papaya mealybug with the parasitoid release. Mulberry is cultivated in about 300 acres in Kerala, mainly in Idukki, Wyanad and Palakkad districts. During 2009-10, mulberry gardens were severely infested by papaya mealybug. Suppression of the mealybug by the parasitoid had saved the silk industry in Kerala. Muberry crop to the tune of 2.1 crores was saved by controlling the mealybug with *A. papayae*. Tapioca in 75000 ha in Kerala, severely damaged by *P. marginatus*, was cleared by release of *A. papayae* in 2011 resulting in the net savings of 2.5 lakhs/ha and 1.8 crores/year.

Infestation of *P.marginatus* was observed on rubber trees and teak nurseries in Kerala. Timely release of the parasitoid had saved these plantation crops without much crop loss. *A. papayae* was found highly effective in suppressing *P. marginatus* on rubber. An area of 5.17 lakhs ha^{-1} is cultivated with rubber in Kerala accounting for 92 per cent of the total area in India with average production of 1949 kg/ha. At an average price of Rs. 16 to 19/kg, considering yield reduction of about 10 per cent, in total infested area of 49500 ha, savings to the tune of Rs. 17.85 lakhs was accrued (Mani Chellappan, 2011).

Orissa

P. marginatus was noticed as serious pest on several agricultural and horticultural crops including papaya in 2011. Biological control programme was initiated in 2011 itself with the release of *A. papayae*. There were no reports of upsurge of papaya mealybug from any part of Orissa.

North East Region

More than 60 per cent of papaya plants were infested with PMB during 2009-10 in four districts of Tripura. Biological control programme was initiated in 2011 with the release of *A. papayae* during January 2011 at Lembucherra. Susequently, there were no reports of incidence of papaya mealybug from Tripura (Agarawala, 2011). *P. marginatus* was noticed in January 2011 in Manipur and *A. papayae* was released in the same month in Manipur. Similar observation was made after the release of *A. papayae* in Assam where there was an incidence during Jan, 2011 in Guwahati and Jorhat.

Lakshadweep Islands

Papaya is the major fruit crop and it was severely affected by mealybug. *A. papayae* was released in Kavaratti and Agathi islands in May, 2011. The parasitoid had established very well in Androth, Kalpeni and other islands and suppressed the pest. In November 2011, severe PMB infestation of papaya leaves and fruits in the Minicoy Island brought down the production quite significantly. The affected trees failed to flower and bear fruits. Three releases of *A. papayae* were made in December 2011, March 2012 and June 2012 in Minicoy resulting in good establishment of parasitoids. Due to the mealybug control, the infested trees got revitalized, rejuvenated and started flowering and bearing the fruits.

Rajasthan, Gujarat and New Delhi

P. marginatus appeared in October, 2012 around Udiapur in Rajasthan, 2011 (Mani *et al.*, 2012) in Gujarat and 2014 in New Delhi. Biological control programme with *A. papayae* was taken up immediately after the appearance of *P. marginatus* in these states.

References

Agarwala B K 2011. A Preliminary Report of the Papaya Mealybug *Paracoccus marginatus* (Hemiptera: Pseudococcidae) in Tripura. Proceedings of the National consulation meeting on strategies for deployment and impact of the imported parasitoids of papaya mealybug, Classical biological control of papaya mealybug (*Paracoccus marginatus*) in India, 30th October 2010, Bangalore, Pp 45-46.

Chandele A G, Nakat R V, Pokharkar D S, Dhane AS and Tamboli N D 2011. Status of papaya mealybug, *Paracoccus marginatus* W and G (Hemiptera: Pseudococcidae) in Maharashtra, Proceedings of the National consulation meeting on strategies for deployment and impact of the imported parasitoids of papaya mealybug, Classical biological control of papaya mealybug (*Paracoccus marginatus*) in India, Pp 43-44.

Dharajothi B, Surulivelu T, Rajan T S and Valarmathi R 2011. First record on the establishment of the parasitoid, *Acerophagus papayae* Noyes and Schauff of papaya mealybug *Paracoccus marginatus* Williams and Granara de Willink) on cotton. *Karnataka Journal of Agricultural Sciences* **24:** 536-537.

Divya S 2012. Studies on management of papaya mealybug, *Paracoccus marginatus* (Williams and Granara de Willink) (Pseudococcidae: Hemiptera). Thesis submitted to the TamilNadu Agricultural University, TNAU, Coimbatore.

Galanihe L D, Jayasundera M U P, Vithana A, Asselaarachchi N and Watson G W 2010. Occurrence, distribution and control of papaya mealybug, *Paracoccus marginatus* (Hemiptera: Pseudococcidae), an invasive alien pest in Sri Lanka. *Tropical Agricultural Research and Extension* **13:** 81-86.

Jacob Mathew 2011. Status of papaya mealybug on rubber in Kerala. Proceedings of the National consulation meeting on strategies for deployment and impact of the imported parasitoids of papaya mealybug, Classical biological control of papaya mealybug (*Paracoccus marginatus*) in India. 60 pp.

Jonathan E I, Karuppuchamy P, Kalyanasundaram M, Suresh S and Mahalingam C A 2011. Status of papaya mealybug in Tamil Nadu and its management. Proceedings of the National consulation meeting on strategies for deployment and impact of the imported parasitoids of papaya mealybug, Classical biological control of papaya mealybug (*Paracoccus marginatus*) in India. Pp 24-33.

Kalyanasundaram M, Karuppuchamy P, D S ivya, Sakthivel P, Rabindra R J and Shylesha 2011. Impact on release of the imported parasitoid *Acerophagus papaya* for the management of papaya mealybug *Paracoccus marginatus* in Tamilnadu. Proceedings of the National consulation meeting on strategies for deployment and impact of the imported parasitoids of Papaya mealybug, Classical biological control of papaya mealybug (*Paracoccus marginatus*) in India. pp. 68-72.

Krishnakumar R and Rajan V P 2009. Record of papaya mealybug *Paracoccus marginatus* infesting mulberry in Kerala. *Insect Environment* **15:** 142.

Krishnamoorthy A and Mani M 2011. Occurrence of papaya mealybug *Paracoccus marginatus* in Karnataka: IIHR perspective Proceedings of the National consulation meeting on strategies for deployment and impact of the imported parasitoids of papaya mealybug, Classical biological control of papaya mealybug (*Paracoccus marginatus*) in India. pp. 37-39.

Krishanamurthy A, Mani M, Gangavisalkshi P N and Gopalakrishna Pillai K 2011. Classical biological control of papaya Mealybug *Paracoccus marginatus* using exotic parasitoid, *Acerophagus papayae*. Proceedings of National symposium on Harnessing Biodiversity for Biological control of Crop pests, abstract. May 25-26, 2011, NBAII, Bangalore. 101 pp.

Lyla K R and Philip B M 2010. Incidence of papaya mealybug *Paracoccus marginatus* Williams and Granara de Willink (Hemiptera: Pseudococcidae) in Kerala. *Insect Environment* **15:** 156.

Mahalingam C A, Suresh S,Subramanian S, Murugesh K A, Mohanraj P and Shanmugam R 2010. Papaya mealybug, *Paracoccus marginatus* - a new pest on mulberry, *Morus* spp. *Karnataka Journal of Agricultural Sciences* **23:** 182-183.

Mani M, Shivaraju C and Sylesha A N 2012.*Paracoccus marginatus*, an invasive mealybug of papaya and its biological control. *Journal of Biological Control* **26:** 201-216.

Mani M, Shylesha A N and Shivaraju C 2012. First report of the invasive papaya mealybug *Paracoccusmarginatus* Williams and Granara de Willink (Homoptera: Pseudococcidae) in Rajasthan. *Pest Management in Horticultural Ecosystem* **18**: 234.

Mani Chellappan 2011. Status of papaya mealybug, *Paracoccus marginatus* Williams and Granara de Willink in Kerala. The Proceedings of the National consulation meeting on strategies for deployment and impact of the imported parasitoids of papaya mealybug, Classical biological control of papaya mealybug (*Paracoccus marginatus*) in India. pp. 40-42.

Meyerdirk D E, Muniappan R, Warkentin R, Bamba J and Reddy G V 2004. Biological control of the papaya mealybug, *Paracoccus marginatus* (Hemiptera: Pseudococcidae) in Guam. *Plant Protection Quarterly* **19**: 110-114.

Miller D R and Miller G L 1999. Notes on a new mealybug (Homoptera: Coccoidea: Pseudococcidae) pest in Florida and the Caribbean: the papaya mealybug, *Paracoccus marginatus*Williams and GranaradeWillink. *Insecta Mundi* **13**: 179-181.

Mundale M and Nakat R 2011. Successful control of papaya mealybug using *Acerophagus papayae* in farmer's field. National symposium on harnessing Biodiversity for biological control of crop pests- abstracts, NBAII, Bangalore, pp. 27.

Muniappan R, Meyerdirk D E, Sengebau F M, Berringer D D and Reddy G V P 2006. Classical biological control of the Papaya mealybug, *Paracoccus marginatus* (Hemiptera: Pseudococcidae) in the Republic of Palau. *Florida Entomologist* **89**: 212-217.

Muniappan R, Shepard B M, Watson G W, Carner G R, Rauf ASartiami D, Hidayat P, Afun J V K and Ziaur Rahman A K M 2009. New records of invasive insects (Hemiptera: Sternorrhyncha) in Southeast Asia and West Africa. *Journal of Agricultural Urban Entomology* **26**: 167-174.

Nakat R V, Pokharkar D S, Dhane A S and Tamboli N D 2011. Biological impact of *Acerophagus papayae* (N and S) on suppression of papaya mealybug *Paracoccus marginatus*(W and G) in Pune region of Maharashtra. Proceedings of the National consulation meeting on strategies for deployment and impact of the imported parasitoids of papaya mealybug, Classical biological control of papaya mealybug (*Paracoccus marginatus*) in India, pp. 79-81.

Pokharkar D S, Nakat R V, Tamboli N D and Dhane A S 2011. Papaya mealybug, *Paracoccus marginatus* Willams and Granare de Willink (Hemiptera: Pseudococcidae) and its natural enemies in Maharashtra. National symposium on harnessing Biodiversity for biological control of crop pests- abstracts, NBAII, Bangalore. 29 pp.

Qadri S M H, Shekhar M A, Vinod Kumar and Narendrakumar J B 2011. An impact and constraints on the establishment of *Acerophagus papayae* for the management of papaya mealybug in mulberry ecosystem. National symposium on harnessing Biodiversity for biological control of crop pests- abstracts, NBAII, Bangalore. 37 pp.

Rabindra R J 2010. NBAII pioneers Successful classical biological control of papaya mealybug. *NBAII News letter***11**: 1.

Regupathy A and Ayyasamy R 2009. Need for generating baseline data for monitoring insecticide resistance in new invasive Mealybug *Paracoccus marginatus* Williams and Granara de Willink (Insecta: Hemiptera: Pseudococcidae), the key pest of papaya and biofuel crop, *Jatropha curcas*. *Resistant Pest Management Newsletter* **19**: 37-40.

Regupathy A and Ayyasamy R 2011. Impact of papaya mealybug *Paracoccus marginatus* on papain industry. Proceedings of the National consulation meeting on strategies for deployment and impact of the imported parasitoids of papaya mealybug, Classical biological control of papaya mealybug (*Paracoccus marginatus*) in India. pp. 57-59.

Sajeev T V 2011. Classical biocontrol of Papaya mealybug *Paracoccus marginatus* (Hemiptera: Pseudococcidae): the forestry perspective. Proceedings of the National consulation meeting on strategies for deployment and impact of the imported parasitoids of papaya mealybug, Classical biological control of papaya mealybug (*Paracoccus marginatus*) in India pp.61-62.

Sakthivel N 2013. Field performance of three exotic parasitoids against papaya mealybug, *Paracoccus marginatus* (Williams and Granara de Willink) infesting cassava in Tamil Nadu. *Journal of Biological Control* **27**: 83-87.

Shylesha A N, Dhanyavathi P N and Shivaraju C 2011a. Mass production of parasitoids for the Classical Biological Control of Papaya mealybug *Paracoccus marginatus*.Proceedings of the National consulation meeting on strategies for deployment and impact of the imported parasitoids of papaya mealybug, Classical biological control of papaya mealybug (*Paracoccus marginatus*) in India. pp. 63-67.

Shylesha A N, Rabindra R J, Shekhar M A, Vinod Kumar, Narendra Kumar and KrishnamurthyA 2011b. Impact of Classical biological control of the papaya mealybug *Paracoccus marginatus* using *Acerophagus papayae* in Karnataka. Proceedings of the National consulation meeting on strategies for deployment and impact of the imported parasitoids of papaya mealybug, Classical biological control of papaya mealybug (*Paracoccus marginatus*) in India. pp. 73-78.

Suresh S Jothimani, Sivasubrmanian R, Karuppuchamy P, Samiyappan P and Jonathan R 2010. Invasive mealybugs of Tamil Nadu and their management. *Karnataka Journal of Agricultural Sciences* **23**: 6-9.

Vijay S 2010. Management of papaya mealybug in cassava (http: //www.thehindu.com/todays-paper/tp-features/tp-sci-tech-and agri/management-of-papaya-mealybug-in-cassava/article464560.ece)

Index